# Wisdom of the Property Crowd

Property industry insight, perspective, and advice from world-class practitioners.

These are the tutors you wish you had in college!

Adam Muggleton

Sold Out Publishing™

Copyright © 2023 Sold Out Publishing™. All world rights reserved.

All rights reserved. No part of this publication may be reproduced, distributed, or transmitted in any form or by any means, including photocopying, recording, or other electronic or mechanical methods, without the prior written permission of the publisher, except in the case of brief quotations embodied in reviews and certain other non-commercial uses permitted by copyright law.

Published by Sold Out Publishing™
Brooklyn, NY 11201
USA

Author: Adam Muggleton

First Printing, 2023

Book cover design by: spncrrbns

Illustrations: jibrahim

Library of Congress Control Number: 2023932555

ISBN: 978-1-7364118-9-6

## Publishers Legal Disclaimer

The author has compiled this publication with care but does not guarantee and assumes no liability for the accuracy or completeness of the information or its suitability for any particular purpose. This publication presents a wide range of opinions on a wide range of topics related to the property industry, specifically the design, construction and facilities maintenance of buildings. These opinions reflect the research, lived experience and ideas of the author and the Edifice Complex podcast interviewees and are not intended to substitute for the professional services of qualified and experienced property industry practitioners. The author, publisher and interviewees of the Edifice Complex podcast shall not be held liable for any losses, injuries, or damages directly or indirectly from the display or use of information, opinions or insights contained within this publication nor from any podcast interview.

No relationship of any description is created or implied between the author and reader/user; nor is any service of any description or advice of any type provided by the author to the reader/user. The entire risk of the use of any information in this publication is assumed by the reader/user. If the reader/user is unclear about this disclaimer or is unwilling to accept the entire risk of using any information in this publication, they should not read the publication.

# TABLE OF CONTENTS

**06**    **AUTHOR**
Adam Muggleton

**08**    **ACKNOWLEDGEMENTS**

**09**    **PURPOSE**

**09**    **USER GUIDE**

**10**    **SAEED AL ABBAR**
Sustainability

**18**    **DR. PETER SIMMONDS**
Tall Buildings & Radiant Systems

**28**    **STEVE BURROWS CBE**
Engineering the Impossible

**38**    **MARCEL HARMON**
Constructing our Niches

**47**    **DR. STEVEN FAWKES**
Finance & Energy Efficiency

**53**    **LLOYD ALTER**
The Case Against Net Zero

# TABLE OF CONTENTS

**62**   **HENRY GORDON-SMITH**
Agritecture

**72**   **DAN NALL**
Design Principles & Discipline

**82**   **GLEN SPRY**
The Coming Energy Pivot

**92**   **DR. ROCHELLE ADE**
Green Building Certification is not Working

**101**   **BILL BROWNING**
Biophilic Design

**112**   **BILL GNERRE**
Why Does it take so long to set up BMS?

**119**   **MURRAY GUY**
Lean Construction

**125**   **PROFESSOR ROLAND CLIFT**
Ethics

**134**   **JERRY YUDELSON**
The God Father of Green

# Adam Muggleton

Adam Muggleton MRICS has been immersed in property and building systems performance for 40 years. Having worked on projects in 21 countries and held leadership positions at several firms, Adam has a unique perspective from convergent experience in property development, design team management, project management, and building commissioning.

**Adam's focus is on property development and delivery as a:**

- Chartered Project Management Surveyor
- Qualified Building Commissioning Professional
- Property Industry Philosopher - www.bldwhisperer.com
- Property Industry Podcaster - www.edificecomplexpodcast.com

The question is, "why are zero defect, high-performance buildings the exception and not business as usual?"

**Author on media platforms:**

Youtube

Twitter

Linkedin

# Acknowledgements

The publisher and author would like to thank all the Edifice Podcast guests who are included in this book. Their work and thought leadership, some of which are distilled within this book, offer valuable insights for anyone working within the property industry.

They are listed here, in no particular order:

- Dan Nall - Design Principles & Discipline
- Dr. Rochelle Ade - Green Building Certification is not Working
- Bill Browning - Biophilic Design
- Henry Gordon-Smith – Agritecture
- Marcel Harmon - Constructing our Niches
- Lloyd Alter - The Case Against Net Zero
- Dr. Peter Simmonds - Pushing the Envelop and Mega Tall Buildings
- Saeed Al Abbar - United Arab Emirates GBC
- Murray Guy - Lean Construction
- Jerry Yudelson - The God Father of Green Reinventing Green Building
- Bill Gnerre - Why Does it take so long to set up BMS?
- Glen Spry - The Coming Energy Pivot
- Professor Roland Clift – Ethics
- Steve Burrows CBE - Engineering the Impossible
- Dr. Steve Fawkes - Finance & Energy Efficiency

Also, a final acknowledgement to ASHRAE Distinguished Lecturer and Edifice Complex podcast co-host Robert Bean. The author recognises that Robert is the calming influence he needs to help navigate podcast interviews and, via his gravitas, attracts high-profile academic guests.

# Purpose

Designing, constructing, and operating buildings is an applied science. It requires large teams with deep domain knowledge and skills. It is like conducting an orchestra with the conductor being the development project manager and the orchestra consisting of all the specialists such as architects, engineers, artisans, tradespersons, contractors, and specialist vendors.

On the worst projects, everyone works deep within their silos and effectively work against each other. With the right leadership, the best projects have everyone working together outside their silos to deliver their best work in the service of the project outcome.

**In my experience, there are two factors necessary for the delivery of excellent project outcomes:**

1. Individuals wanting to do their best work and "bring it" every day
2. Leadership that inspires great work and can produce the 2+2=5 effect

This book is not about leadership, but the people interviewed on the Edifice Complex podcast are leaders in their field. This book is for individuals wanting to do their best work every day and develop their careers by studying role models applying best practices.

For people early in their careers, training or just seeking to be inspired, role models are, in my opinion, necessary for personal development. We need to see what is possible and what excellence looks like, we need people to model ourselves on.

The purpose of this book is to distil some of the critical thinking, insight, and ideas of several of the property industries' most accomplished and respected practitioners. I have always been frustrated with the property industry's ability to "hide in plain sight" their best, most innovative, and experienced practitioners. This book aims to suggest role models and highlight great work based on my lessons learned during the interview process.

# User Guide

I like books that are easy to navigate and get to the point. I have tried to do that here by keeping the chapters brief and broken down into sub-headings that allow the reader to decide if they want to read further. Space is provided for the readers to take notes at the end of each chapter and the book's end.

The order of the chapters was something I contemplated a lot, but, in the end, I opted to organize them at random because, in my opinion, each chapter stands on its own. Therefore, this is a chose your own adventure book. I encourage the reader to review the chapters and dive into the ones that arouse your curiosity. I hope you enjoy the ride and find inspiration.

# Saeed Al Abbar
Sustainability

Saeed is that unusual combination of engineer, leader and entrepreneur in a business sector that is notoriously slow to change and tends to ignore young thought leaders. Somehow Saeed has managed to cut through the noise and barriers put in front of him and create an award-winning company that is an excellent example of gender inclusion in a traditionally male-oriented industry.

Of course, he has not done this independently; Saeed has partners, employees, and a support team around him. All consulting firms are the total of their people's talent and energy. Consulting firms succeed when their leadership recruits and empowers talented people without micromanaging them. Leaders succeed through the empowerment and coordinated actions of others. AESG is a great example of a forward-looking, progressive firm that platforms its thought leaders, safe in the knowledge that their success is everyone's success.

Scan the QR code to view Saeed Al Abbar's Linkedin Profile

When we interviewed Saeed for the podcast, he was also chairman of the Emirates Green Building Council and active in the World Green Building Council. His advocacy, insights, and ability to think at the macro and micro levels were impressive. Early in their career, any built environment professional should consider Saeed a role model for what is possible.

Saeed Al Abbar is a founder and CEO at AESG in Dubai, UAE. He has been actively involved in managing and directing some of the world's most prestigious projects through design, construction, commissioning, and operation. His consultancy and commissioning services span through the Middle East, Europe, and Asia.

An author of several papers, he has also presented his work at various local, regional, and international conferences. In addition to his duties as CEO of AESG, Saeed is also the Chairman of the Emirates Green Building Council and a Board Member of the World Green Building Council. He plays an active role in promoting and advancing sustainable building designs in the UAE and MENA region. Saeed is the recipient of the International Young Consultant of the Year award by the British Expertise International Awards in the UK. In addition, he has also been named in Gulf Business' Top Young Achievers "30 under 30".

*"Not that we've given up, but I think that there's a realisation that the naivety of trying to change things by just getting people to do things out of the goodness of their heart is not going to work."*

Figure 1 MC2 - Masdar City Development, UAE. Source: AESG

## The Current State of Play

Many buildings are built speculatively with no certainty about who will occupy them or even what business activities will be undertaken. Consequently, this leads to generic "copy and paste" design solutions, traditional inefficient construction methods and poor occupant experiences. Overall, this scenario advances mediocrity. In extreme cases, this approach can lead to empty buildings or, from an investor's perspective, "stranded assets."

Considering the vast sums of money involved in designing and constructing a building, this is embarrassing and a poor allocation of resources. Furthermore, in a world of growing awareness about the externalities of the built environment and the need for sustainability, pressure is mounting for real change.

## Leadership and Change

Saeed believes that change in the property industry will be driven from local and city levels and not necessarily by national policies. This matters because, cities are significant generators of GDP and the economic engines of nation-states.

Due to high population densities and high levels of autonomy, cities are the worst hit by climate change and environmental degradation. Accordingly, they have been leaders in tailoring policy to mitigate these effects and attract financial plus human capital. Cities are likely to be the "canaries in the coal mine" and consequentially leaders in mitigation legislation due to ever-increasing urbanisation

Scan the QR code to listen to Saeed Al Abbar's Edifice Complex Episode

# Micro-level Engagement

One way to influence sustainability is to create systems that get the people involved at the lowest levels without necessarily joining any companies or businesses. This can be achieved by leveraging technology to mobilise people and provide jobs that are in alignment with future trends.

There is a macro and micro level to engagement and reducing environmental degradation. At the macro-level, policy needs to be developed with an eye to forecasting and future trends. At the micro-level, beneficial behaviour could be encouraged via "nudging" people with novel systems and incentives.

Gamifying systems with rewards via apps that encourage recycling and reflection on what to and what not to purchase can generate new, environmentally beneficial economic models, similar to how gaming apps have created wealth and transfer of money. The new micro-economy is driven by technology that benefits the environment and reduces externalities in an ideal world.

Figure 2 ICD Brookfield Place, Dubai UAE. Source: AESG

Wisdom of the Property Crowd | 13

# The Challenge for the Global Green Building Movement

Figure 3 Saeed Presenting at an Industry Conference. Source: AESG

Thanks to the 2015 Paris Climate Accord, there is now increased public realisation on the need and acceptance for change. Consequently, there seems to be a growing cultural awareness that the status quo is not sustainable. Externalities from GHG emissions are acute and are everyone's and no one's problem at the same time which is a true tragedy of the commons dilema.

The built environment must transform from showcase and vanity projects to demonstrably sustainable projects. For example, several countries are targeting all new buildings to be NetZero by 2030. This is a real change for the better, in my opinion. However, the newly built property sector is only approximately 1% of the total building stock. Therefore, the real issue is what to do with the other 99%, i.e., the existing building stock?

New York City in the USA has addressed this issue with legislation to incrementally reduce carbon emissions to achieve a 40% reduction by 2030 from a 2005 baseline and 80% by 2050. This is a promising move, and other cities should follow this example. New York City realised that legislation is required because the cost of energy is too low to act as an incentive to modify buildings and reduce GHG. Sometimes a stick i.e., fines and a carrot i.e., tax relief are required. Each city has to find the right level of "carrot and stick."

It is inspiring that city leaders and people are responding to such an important challenge. Cities and local authorities are signing up and pledging commitment to 2030 and 2050 targets. In my opinion, change is actually coming from the city level because local politicians are more in touch with their constituencies and local economies.

## Building Codes & Regulations

National and city building codes & regulations typically lag industry and culture by about ten years. This implies that they seem to be perpetually behind current trends and needs. This is a tragedy, given that all the building codes lead to minimum compliant designs. In practice, they become design maximums, not design minimums. They do not encourage any innovation or extension of the performance envelope.

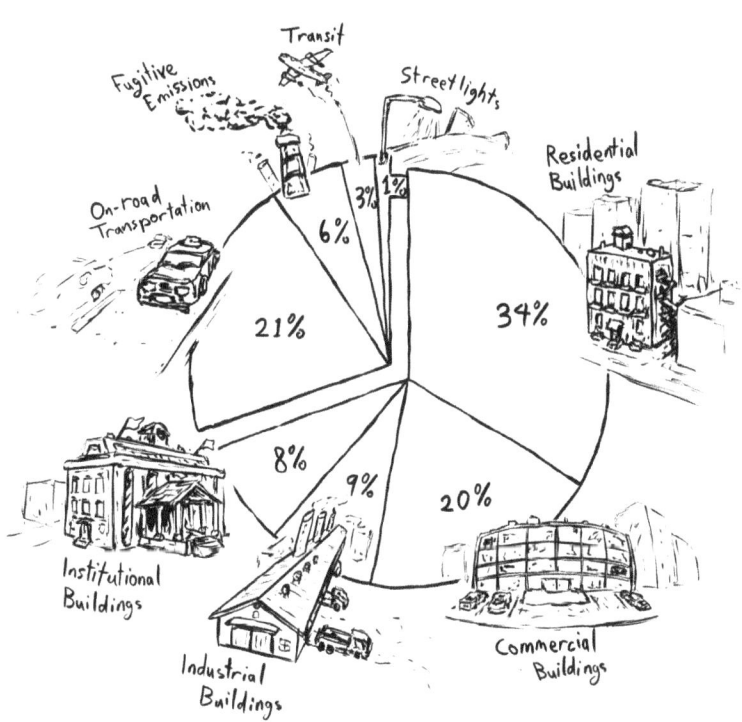

Figure 4 New York City 2013 Greenhouse Gas Emissions by Sector

The bottom line is that building codes & regulations need to be simplified or maybe performance-based. Moreover, they also need to be updated frequently and become aggressive, i.e., set energy and sustainability targets that exceed current practice. The cadence of change needs to increase to meet 2030 and NetZero targets. Therefore, we must set the targets via legislation and building codes and then let the market provide the solutions.

## Gender Inclusion Within the Property Industry

For Saeed, any organisation within the property industry, which does not embrace gender inclusion, stands to miss many advantages. Saeed strongly believes that there is a lot that women have to offer in the property industry as a whole. Using his own organisation as a point of reference, having 42% female staff strength and half of the AESG management team being female, he believes that many of the issues holding back progress in the built enviroment are fueled by male egos.

## Reconnecting Building Design and Construction

Saeed is not a fan of architects, designers, and engineers who do not spend time on-site to follow up on their projects during construction. The common practice of the design firm field inspectors cleaning up architects, designers, and engineers mess on site is low value and, frankly, disrespectful to the client.

With architects, designers, and engineers following through with site inspections, a real learning and feedback process can take place. Learning from real-world on-site issues is a self-improvement learning loop. A mistake discovered and corrected should not be repeated in future projects.

Scan the QR code to visit AESG.com

# Reader Notes: Sustainability

Readers are encouraged to note their favourite insights, triggered ideas, and next steps.

# Dr. Peter Simmonds
## Tall Buildings & Radiant Systems

Dr. Peter Simmonds is, in my opinion, one of the top 10 building services engineers in the world with regards to Tall, Super Tall and Megatall buildings. He literally wrote the ASHRAE book on designing building services systems for skyscrapers. As well as being a leader in skyscraper design, Peter is also an acknowledged thought leader and design engineer for natural ventilation plus radiant heating and cooling systems.

Dr. Peter Simmonds is an academic. However, he is also a practising engineer, with a portfolio of completed projects that are working examples of his academic work translated into reality. In short, Peter talks the talk and walks the walk.

We live in a world where architects and developers win glossy awards and engineers play the supporting role. This leads to a situation where school children and young graduates lack visible engineering role models. We were keen to interview Peter on the podcast because he is an excellent example of an engineer's impact on the world. Peter is, in my opinion, a fantastic role model for school children and young graduates to emulate.

**"A tall building isn't just one floor multiplied by 150 floors. It's lots of other things and how they all talk to each other."**

Dr. Peter Simmonds has attained two Bachelor of Science degrees, one in Mechanical Engineering and the other in Research and Development, and also has a Masters degree and a Ph.D.

Peter is an ASHRAE Fellow Life Member, he received the, The Crosby Field Award (2019) and the John James International Award (2020) from ASHRAE. He has also been recognized with the ASHRAE Distinguished Service and Exceptional Service awards. Peter has also received the Carter Bronze Medal from the Chartered Institution of Building. Moreover, he is regarded as a recognized authority in the field of radiant heating and cooling systems, as well as thermal comfort. He has authored and co-authored more than 60 technical papers, articles, and books. In addition, Peter has written a book on designing skyscrapers, he is the author of the ASHRAE Design Guide for Tall, Supertall, and Megatall Building systems. Also, together with Erin McConahey, Peter co-authored the ASHRAE Design Guide for Natural Ventilation.

## The Diminishing Relevance of LEED

Like every other industry, the construction industry will have to step up, both in methods and in practice, as technology advances and environmental issues become more prominent. LEED is suffering a side effect of construction industry advancement, mainly because it lacks a tangible numerical value, i.e., it is vague on qualifying and publishing performance data. When dealing with high-rise buildings, we inevitably talk about a lot of numbers as well as the associated data, huge numbers running into millions. Therefore, LEED is becoming irrelevant, and, in Peter's opinion, it will fade away.

Scan the QR code to view
Dr. Peter Simmond's Linkedin Profile

Figure 5 Cooper Union, New York, a LEED Platinum building

## Net Zero and Realistic Energy Conservation Goals

While Net Zero and Carbon Neutral are currently fashionable, few people understand the definitions and reality. While striving to achieve carbon-neutral is a lofty goal, most people do not understand that there are several limitations. It is almost impossible to cut the carbon footprint of any occupied buildings completely. What is much more realistic is setting a design goal of an Energy Use Intensity (EUI) based on benchmarking best practices? Buildings that people and machines occupy will obviously consume energy. The key is to reduce, reclaim, absorb and generate energy by design. A practical example of this scenario is the Pearl River Tower building in Guangzhou, China. This building is one of the most environmentally friendly, Net Zero buildings in the world.

## Dr. Peter Simmonds Book on Tall Buildings

Dr. Peter Simmonds' book titled *"The ASHRAE Design Guide for Tall, Supertall, and Megatall Building systems"* is currently in its 2nd edition. The dynamism experienced within the design industry for tall buildings led to the need to update some information in the second edition, which was published in 2020.

The 2nd edition has updated information on how climate effects floor-by-floor systems, and it discusses the complexity of designing buildings in an easy to assimilate form. The book is a guide to help understand the progression of buildings classification from tall to supertall to megatall buildings.

## Climate Considerations for Tall Building Design

With tall buildings, there has to be a serious consideration of the climate change that occurs as you go higher up the floors. Naturally, there is a temperature drop leading to temperature differences across various levels or floors of the building. This means that the building's systems have to be designed and built to consider variable load calculations. Variability in environmental and, therefore, load conditions due to height and complications in ascertaining occupancy rates also make diversity calculations and assumptions essential and challenging. **"Rule of thumb" design calculations do not work designing tall buildings.**

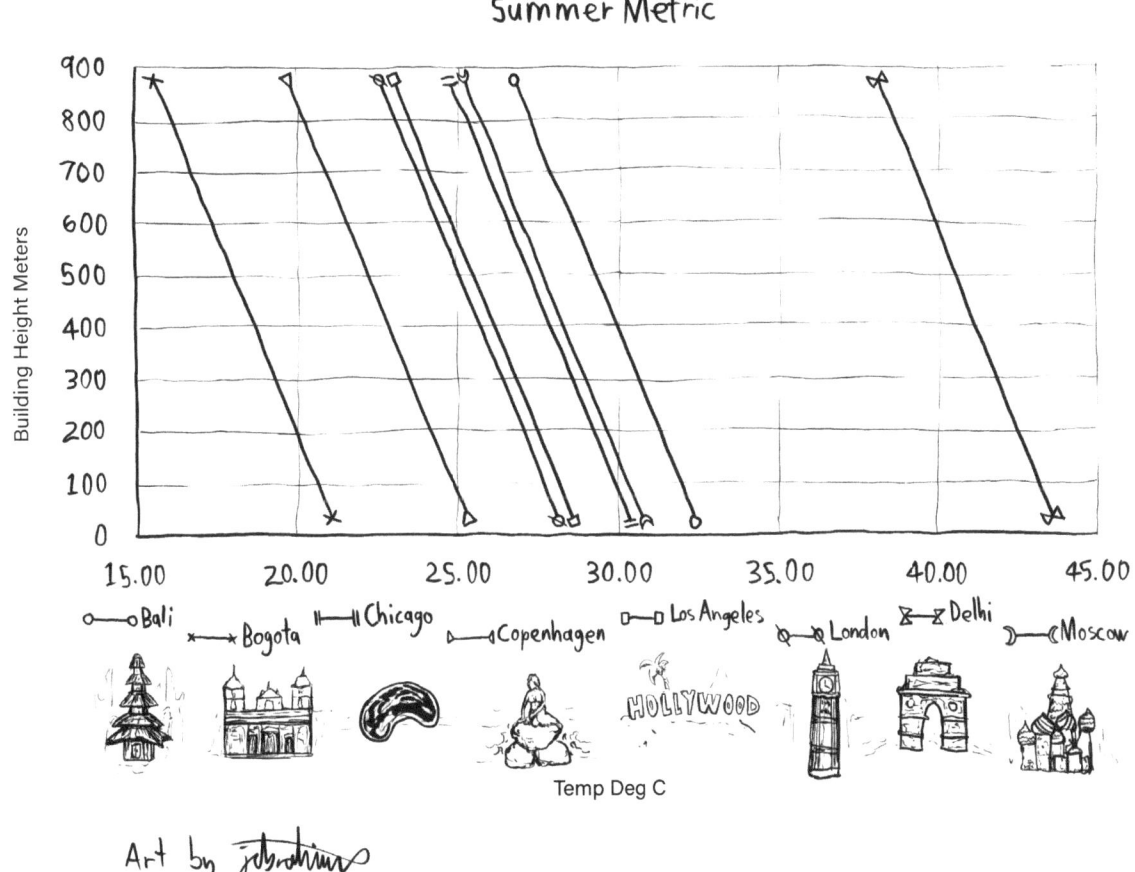

Figure 6 How Outside Temperatures Vary Over the Height of a Mega Tall Building (Summer Time). Data Source: Dr. Peter Simmonds

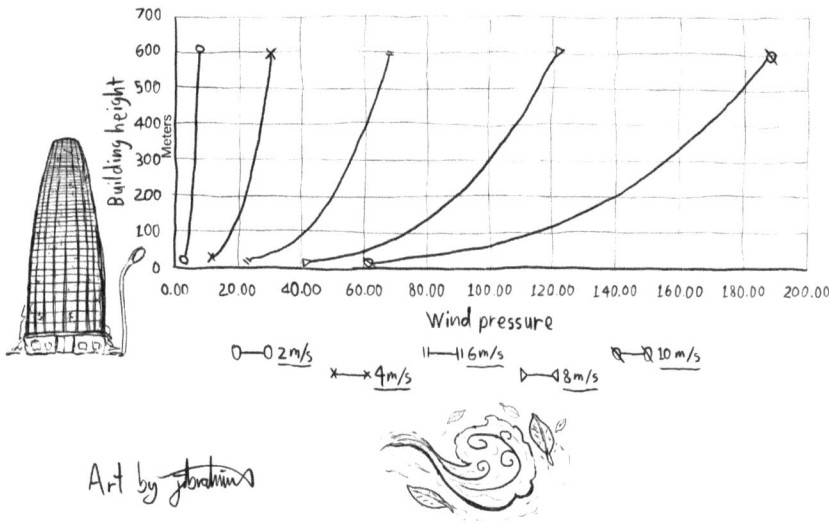

Figure 7 Increase of Wind Pressure Over the Height of a Mega Tall Building. Data Source: Dr. Peter Simmonds

# The Myth of Intelligent Buildings

Dr. Simmonds notes that precise building systems control via real-time, digital measurements and feedback has several benefits. However, there is a mythology of intelligent buildings, which conflicts with precise building systems control.

Precise building control requires minimizing variables wherever possible. However, if you have several people in a space, each with a different level of desired temperature, ventilation, and lighting to be comfortable, one of two scenarios will happen. First, stable temperature, humidity, ventilation and lighting levels will be provided, and some people will be happy and some will not. Second, the systems with smart and/or local user control will be provided, and the user will adjust system set points and performance, creating possible sub-optimal feedback loops. In this case, what happens to your smart space and building?

Smart buildings are a triumph of marketing over substance. Persistence of performance, i.e., stability in environmental conditions, is more desirable and environmentally friendly than perpetual system setpoint changes.

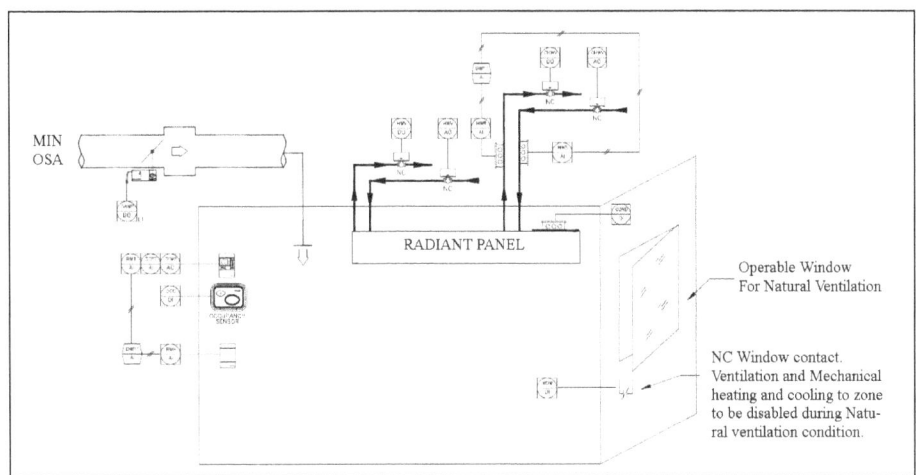

Figure 8 Shows an Example of the Level of Controls Required for Smart Control of a Building Space.
Source: Dr. Peter Simmonds

## Value Engineering

While it is a popular notion that manufacturing is much cheaper in places like China and other Asian regions, there is also an unspoken truth. In a bid to maintain some pseudo affordability, there is also a deliberate declination in the quality of manufactured products. This lowering in quality to present affordability is present in the construction industry at scale. It manifests in low-quality manufactured equipment plus in value engineering during the design process.

Value engineering is hazardous because poor value engineering decisions influence building systems performance over the entire building life span. In Dr. Simmonds experience, value engineering attracts many failed engineers whose primary purpose is to cut costs and make things cheaper at the expense of the building's long-term persistence of performance.

Scan the QR code to listen to Dr. Peter Simmonds's Edifice Complex Episode

# Making Things Work for Architects

It is essential that young Mechanical engineers understand that their often, unappreciated job is to "make things work for architects." Mechanical engineers work to ensure building systems performance by working out building physics and getting them to align with prescriptive building codes. They need to focus on strategies and building performance in addition to meeting the basic building code. Moreover, they also need to be able to say no to architects when necessary by employing full technical reasoning and providing an alternative solution. This is what engineering is all about.

# Radiant Floor & Displacement Ventilation

Typically, a vertical closed-loop ground heat exchanger is used for radiant floor heating and cooling systems. A displacement ventilation system is connected to a cool tube system intalled in a underground pathway. The radiant floor systems and the displacement ventilation system are designed to maintain a constant level of thermal comfort in the occupued zone.

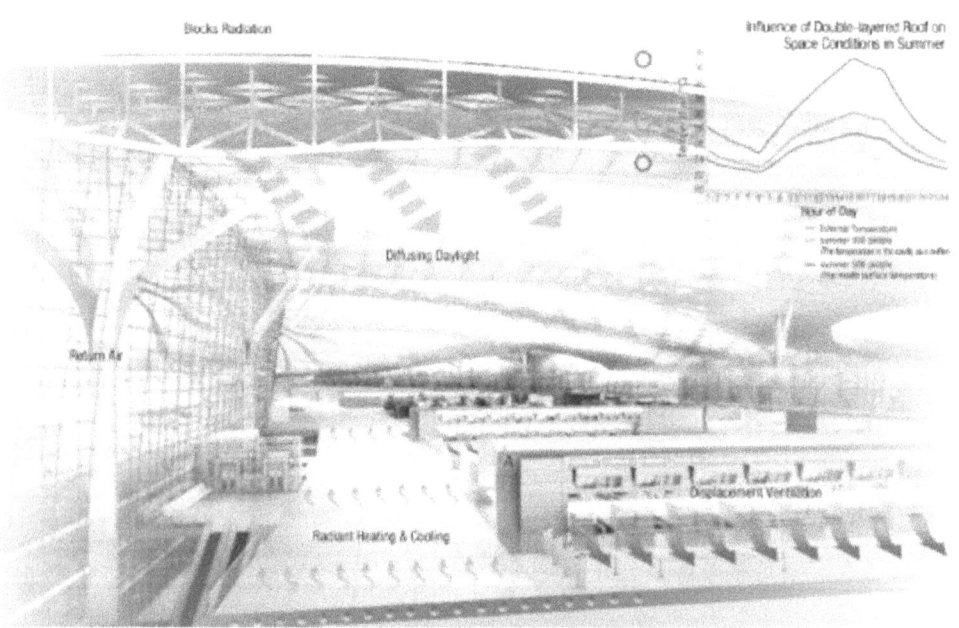

Figure 9 Concept Design for a Large International Airport. Source: Dr. Peter Simmonds

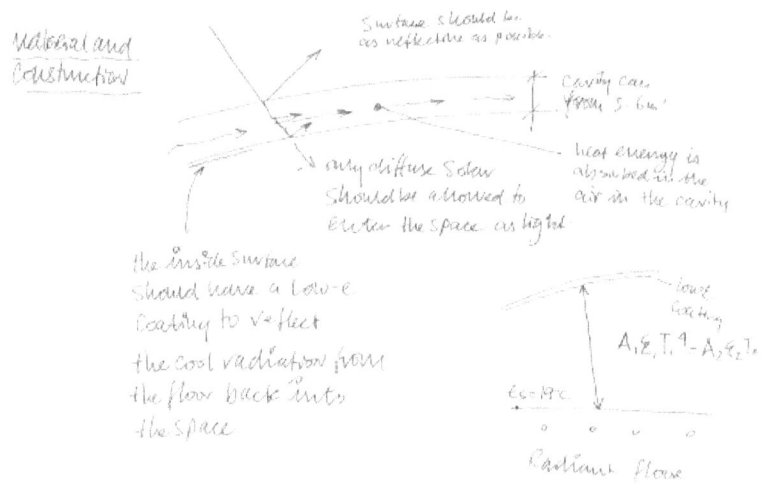

Figure 10 Concept Diagram on How a Building and its Systems Need to Harmonize Each Other. Source: Dr. Peter Simmonds

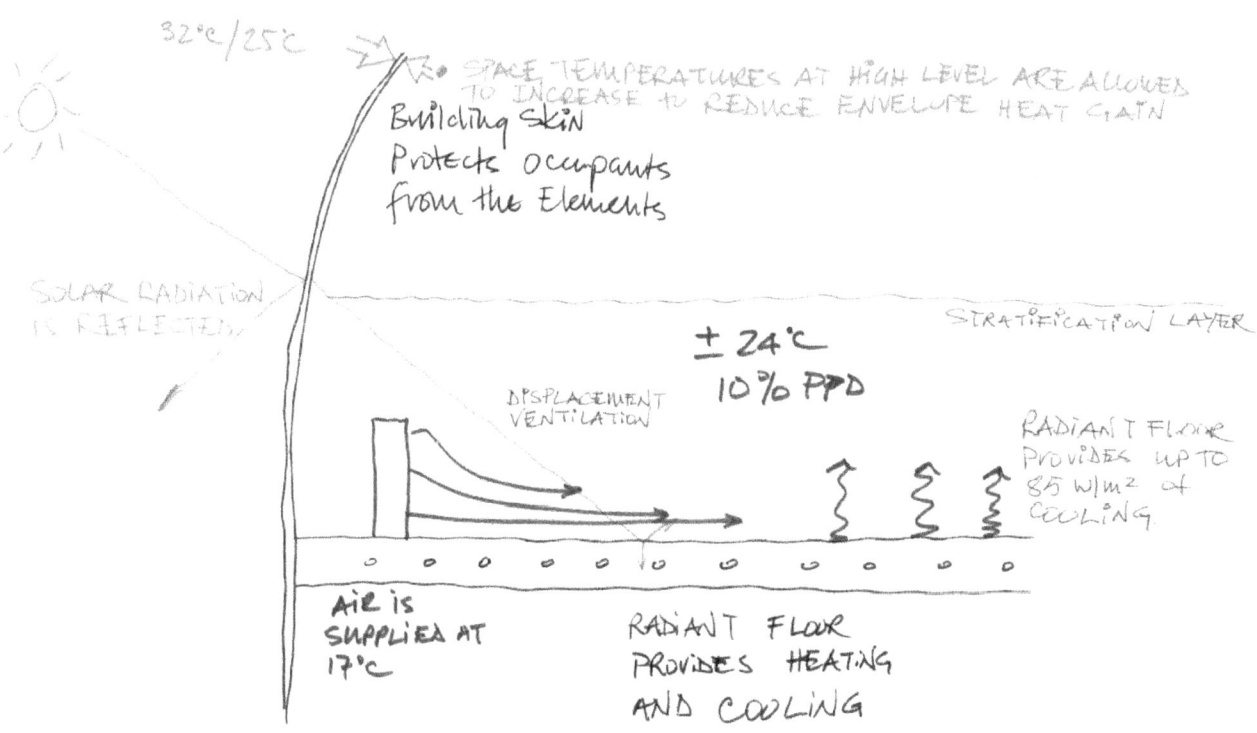

Figure 11 Concept Diagram of the Air Conditioing System Utilizing Displacemnet Ventilation and Radiant Floors for Cooling. Source: Dr. Peter Simmonds

# Bangkok Airport Project

Bangkok airport and its radiant cooling systems designed by Dr. Simmonds stand as a rebuttal of the often stated *"radiant cooling can never work in a tropical climate."* Bangkok airport was a landmark project, not just because of its scale and foot fall but also because it implemented an efficient radiant cooling system.

Several design challenges at the airport included high ceiling heights, high levels of glazing, high humidity climate, and high metabolism occupancy (nervous and excited passengers in large numbers). The final building systems design solution encompassed a strategy of indoor air stratification, i.e., maintaining design conditions in the usable 2.5 meters height and then letting the air temperature increase at the higher levels, high performance envelops incorporating membrane and fritted glass plus full-floor radiant cooling with displacement ventilation. Moreover, CFD was also utilized to test and refine the design solutions. Air systems were used to deal with latent gains and trim conditions in high occupancy regions and entrances. Bottom line, the design soloution worked!

# North America and the Resistance to Change

As opposed to the rest of the world, there is a considerable resistance to transformation in North America, especially regarding moving to dedicated outside air systems and radiant heating & cooling systems. The resistance to change is also impacted by a lack of subject matter experts that understand what they are doing.

Dr. Simmonds' time in Holland designing and implementing radiant systems for hospitals is a testament to the fact that these systems actually work on a large scale. However, due to insufficient knowledge, expertise and corner-cutting in radiant systems in North America, the few attempted systems the date, have high failure rates.

The lack of change and adoption of efficient radiant heating and cooling systems in North America is driven by a design and manufacturer path dependency plus the fear of risk taking to avoid possible liabilities and legal actions. Currently, construction supply chains drive design decisions in North America. Consequently, we are currently building for architects egos and manufacturers preferences. However, we must change and build for people and environmental comfort.

Scan the QR code to visit
petersimmonds.com

# Reader Notes: Tall Buildings & Radiant Systems

Readers are encouraged to note their favourite insights, triggered ideas, and next steps.

# Steve Burrows CBE

Engineering the Impossible

If anyone tells you being an engineer is a boring profession, you should point them towards Steve Burrows. Steve's journey has taken him from school leaver to international award winning engineer, TV documentary host, C Suite advisor, public speaker and CBE (if you are British, you will understand this one).

Steve is enthusiastic about modular construction methods and the intersection of technology and building delivery. He is open minded and forward thinking. Currently, I consider Steve to be the property industry Obi-Wan Kenobi.

Personally, I find Steve's story and achievements very inspiring. He is a great engineer, a macro thinker and one of the most enthusiastic, uplifting people I have interviewed. If I could, I would have Steve visit engineering schools and universities to give motivational talks to students. If Steve does not inspire you as an engineer, engineering might not be for you.

**"You have to have continuous professional development; you have to learn something every year and prove that you've learned something. Success is all about teamwork and never about the lone genius."**

Steve Burrows' long career in civil Engineering has spanned decades. Since graduating from Liverpool Polytechnic in 1982, he worked for the leading engineering firm ARUP, where he was heavily involved in their work on the famous Bird's Nest stadium in Beijing. Moreover, he was made a Commander of the British Empire (CBE) in recognition of his work in engineering worldwide. In addition, he is also a Fellow of the Institution of Civil Engineers, a Chartered Engineer, a Professional Engineer (PE), a member of the Institution of Structured Engineers and a LEED Accredited Professional.

After a long career at ARUP, Steve left to join the US consultancy practice AECOM and then WSP, where he held senior positions. Steve currently works with his own firm, B2Burrows, as well as investing in, and advising, start-ups in the "PropTech" world.

Figure 12 Birds Nest Stadium, Bejing China

Scan the QR code to listen to Steve Burrows' Edifice Complex Episode

Although once afraid of public speaking, Steve has made himself very visible (or audible) on Television and Radio, in keynote addresses, and even in an IMAX movie designed to attract children to STEM subjects. In addition, Steve has also been hosting television documentaries that explore famous archaeological monuments and uncover the engineering secrets that made them possible. He believes that it is the most significant time in history to be an engineer and that optimism carries over into everything he does for his clients.

## The Route Less Travelled

Steve's journey to Chartered Engineer and Fellow of the Institution of Civil Engineers was not very straightforward or predictable. He describes his experience as *"stumbling into Engineering."*

His journey started with him dropping out of high school, going to night school, and finally graduating from Liverpool Polytechnic. Steve chose the Civil Engineering profession because he had a friend who was also studying the same subject. He had no mentor or role models to help him and didn't really know what a Civil Engineer did when he started his undergraduate degree course.

On reflection, Steve believes the path to becoming an engineer should be deliberate and guided. No one should have to stumble into Engineering as he had. In addition, studying at Polytechnic with its practical and applied approach to learning was, in hindsight, an advantage to Steve. It left him with a thirst for continual knowledge that could be applied in his day-to-day engineering practices.

His Alma Mater presented him with a Honorary Fellowship in 2019 and Steve chose not to explain that it was all down to chance when he gave his acceptance speech.

# Modular and Prefabricated Construction (North American Context)

Due to continual increases in land, materials, compliance, and labour costs, designing and constructing a building has become cost-prohibitive to the point where the cost of creating buildings has outstripped GDP over many years, meaning that the lowest 20% of workers cannot afford housing within an hour of their work. Add to this dilemma low construction worker productivity (only 57% on average or, over two days per week wasted) plus decreasing quality outcomes. These issues are compounded due to a looming skills shortage with "baby boomers" starting to retire on mass and too few young trainees coming into the construction industry to be trained. Given all this, modular and prefabrication may be the solution the construction industry has to adopt.

## Overview of Productivity Improvement Over Time

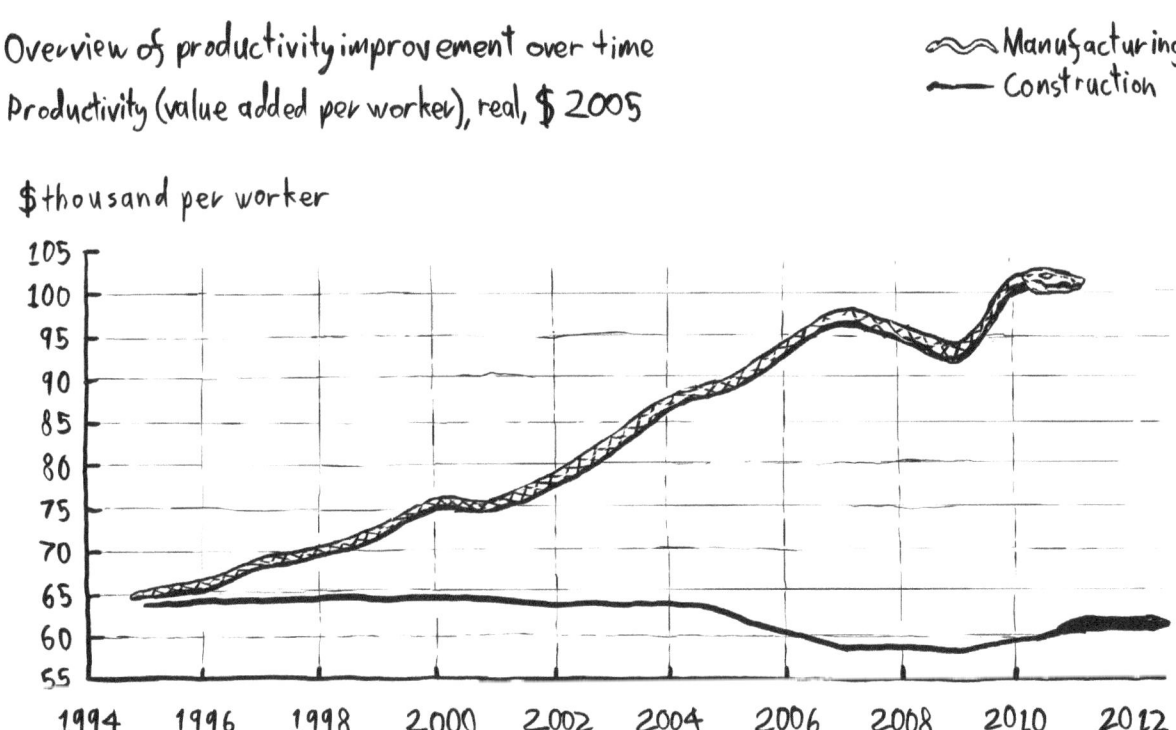

Figure 13  Productivity Improvement Over Time

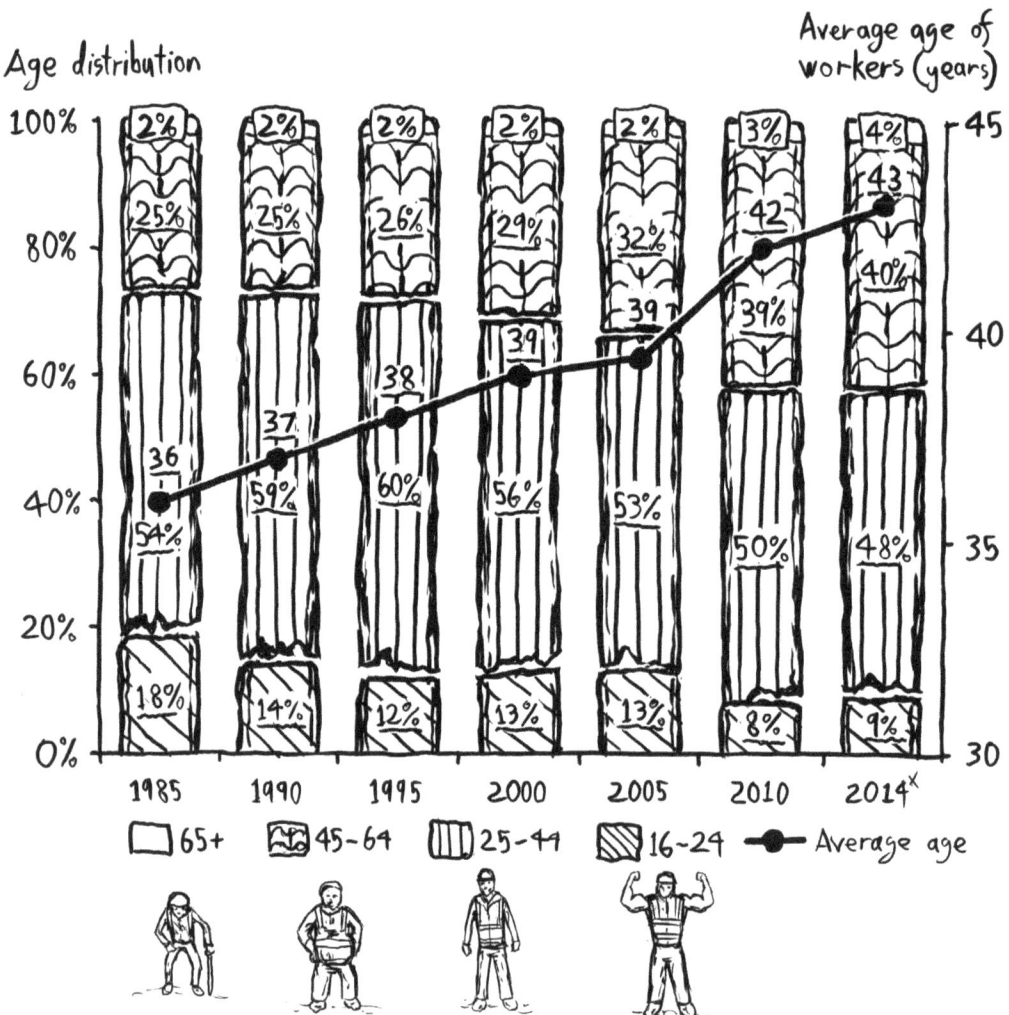

Figure 14 Demographing Within the Construction Industry (USA)

Consider this, the current average age of a carpenter in the USA is 50 years old and rising annually!

To make matters worse the climate crisis demands we reduce carbon emissions. If cement was a country, it would be the 3rd largest carbon emitter in the world, that's the mountain to be climbed.

This current formula of costs rising more than GDP, low productivity, skills shortages, and low quality is not sustainable. Something has to change.

To make building design and construction cheaper and of the quality needed to create energy efficient spaces, a level of standardized processes plus automation is required. This is known as offsite construction which is the interface of technology, robotics and manufacturing and it will, in my opinion, revolutionised the building industry over the next ten years. Steve is currently working with several firms to help bring affordability to the built environment while driving the costs down.

Prefabricated construction and offsite manufacturing provide the following advantages:

- Controlled construction conditions, e.g., fewer bad weather days and better, safer working conditions.
- Improved quality control and new 'greener' materials.
- Greener, more sustainable construction practices, e.g., less waste, more recycling, improved IAQ and VOC reduction.
- Innovative technology opportunities with digital twins, automation, CAD to CAM plus new construction techniques like 3D printing, machine learning, and reusable components.
- Faster construction with predictable completion deadlines with 10% of current site labor.

The construction industry is notoriously slow to change, and any shift at scale to prefabricated construction and offsite manufacturing will require agreement with unions and professional licensing bodies. However, I believe the push will come from governments, codes and agencies as demand exceeds supply in the foreseeable future.

## Building Performance Monitoring

With the advent of the Internet of Things (IoT) and 5G technology, the process of monitoring building operations and performance will become automated, and costs will fall dramatically over time. Steve's rule of thumb is that if a building costs one unit to create, it costs five more units to operate throughout its lifespan. This is the "elephant in the room," the inconvenient truth that few people want to address. Building total lifetime costs post-occupancy often eclipse the design and construction costs.

The automation process of building operations and quantification of performance metrics represent a new opportunity to improve occupants' well-being and reduce building lifetime costs through optimization. Measuring, monitoring and controlling in real-time, indoor air quality, volatile organic compounds, temperature, humidity, occupancy, user trends, lighting levels, window blinds etc., will provide optimization benefits, which have been previously unavailable.

IoT and 5G technology will make buildings more organic and dynamic. This is going to revolutionize facilities management and performance accountability.

Buildings have a relatively short life span, most do not last more than 50 years because they are no longer economically viable. 30% of buildings presently end up in landfill after just 30 years. What if we could disassemble them and reuse them elsewhere?

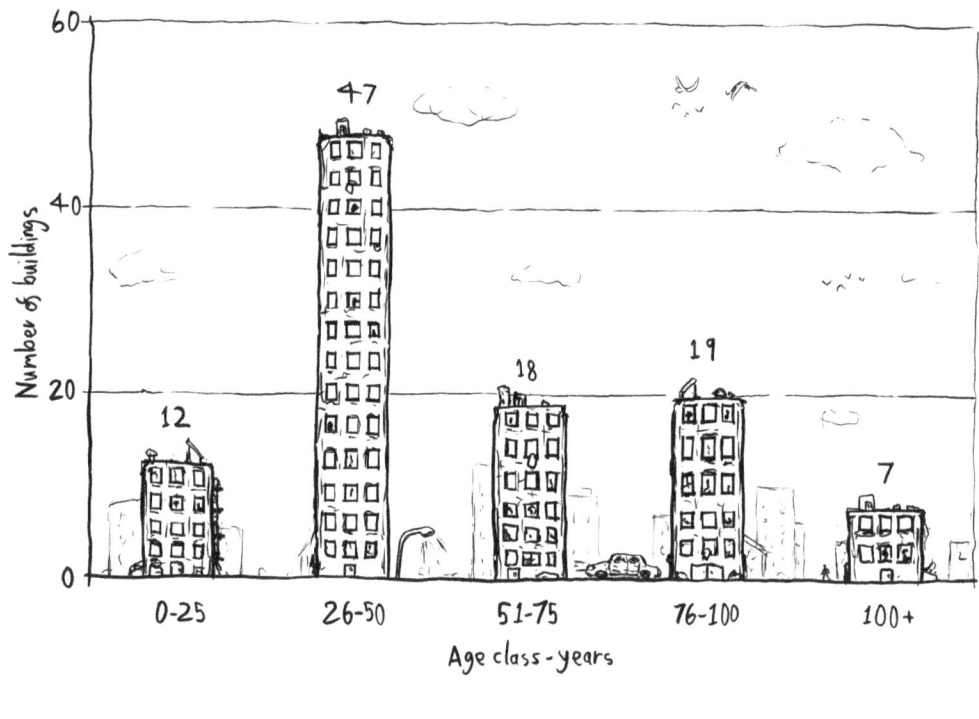

Figure 15 Building Demographics (USA). Data Source: Steve Burrows

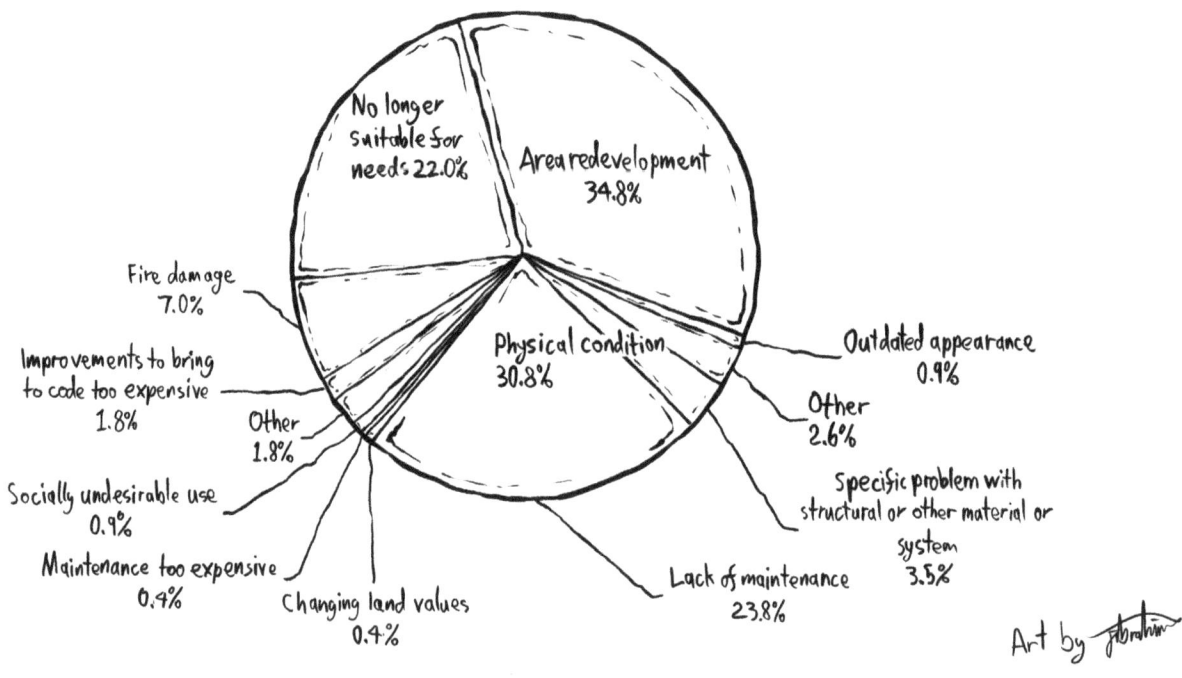

Figure 16 Reasons for Demolition & Redevelopment (USA). Data Source: Steve Burrows

# Percentage of People Dissatisfied

The development of more organic and dynamic buildings requires full consideration of occupants' thermal comfort and satisfaction. The environmental and personal factors that influence individual thermal comfort are summarized in the diagram below.

The engineers working in building services design should consider the standards related to thermal comfort defined in ASHRAE 55. This requires deriving Predictive Mean Vote (PMV), which is an index that aims to predict the mean value of votes of a group of occupants on a seven-point thermal sensation scale. To derive PMV, the simulated temperature and airspeed velocity of a given environment are utilized as input variables. These variables, along with given inputs for clothing insulation, relative humidity, and mean radiative temperature, provide the basis to calculate PMV.

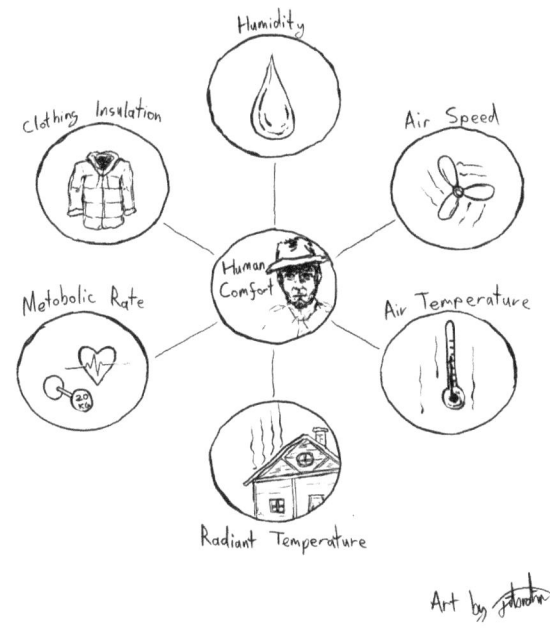

Figure 17 Environmental and Personal Factors that Influence Thermal Comfort. Source: simscale.com/blog

PMV can predict the thermal sensation of a population, and then design engineers need to consider the level of satisfaction of the occupants in a space or building. To do this, they need to derive the Percentage of People who are Dissatisfied (i.e., PPD gives the percentage of people predicted to experience local discomfort, i.e., thermally-dissatisfied occupants). The primary factors causing local discomfort are unwanted cooling or heating of an occupant's body.

The ASHRAE 55 standard states that thermal comfort can be achieved based on =/> 80% occupant satisfaction rate. The remaining percentage of people can still experience 10% dissatisfaction based on whole-body discomfort and 10% dissatisfaction based on local/partial body discomfort. To comply with ASHRAE 55, the recommended thermal limit on the 7-point scale of PMV is between -0.5 and 0.5.

The PPD can range from 5% to 100%, depending on the calculated PMV. These comfort values vary depending upon where the occupant is located in the building. For comfort ranges to comply with the standards, no occupied point in space should be above 20% PPD.

Steve believes that with IoT and 5G technologies, we now have the technology to measure and control these critical comfort factors and increase occupant satisfaction and productivity. Consequently, we are moving into a brave new world where we design for people and not just money.

Scan the QR code to visit Steve Burrows' website

Figure 18 Steve in Jordan for his Time Scanners T.V. Show

## Engineering

Steve is a great example of what is possible in an engineering career. He is a great role model for all aspiring engineers. Moreover, he is also one of the most upbeat and enthusiastic people I have had the great fortune to interview. During our interview, Steve made me smile several times, and I noted down some of his assertions that I think are great takeaways:

*"I think there are more opportunities for women in engineering today than ever in history"*

*"I think the traditional route into engineering is no longer essential"*

*"I think that engineering is much more welcoming of a whole range of skills not only math and science as we think more about people and places and nature and happiness"*

*"Engineering builds deep bonds of friendship. When you do a project together, you've been through an experience, like no other"*

*"The Internet of Things is going to create the opportunity for buildings to be much more dynamic and organic, technology enables the inanimate to come to life"*

# Reader Notes: Engineering the Impossible

Readers are encouraged to note their favourite insights, triggered ideas, and next steps.

# Marcel Harmon

Constructing our Niches

In my experience, building design and delivery have a focus on the engineering and quantitative rather than qualitative influences and user experience.

Given that exceptionally successful businesses such as Apple have a focus on the user experience, I am curious why building design does not have a greater focus on user experience.

Marcel Harmon thinks differently. He is an engineer with knowledge and expertise in Anthropology, and it is an understatement to say that this is not a skill set you encounter every day. Marcel brings evolutionary theory to building architecture, engineering, and construction and, in my opinion, is a great example of a multidiscipline thinker, engineer and solutions provider.

"**cross-disciplinary work breeds innovation…and evolution kind of happened at the boundaries of ecosystems. It's the same for cross-disciplinary work; boundaries across disciplines are where cool things can happen.**"

Marcel Harmon is a licensed Professional Engineer with a Master's Degree and Doctorate in Anthropology. In addition, he is also Associate Principal at Branch Pattern, where he supports the firm's high-performance building service work and also undertakes research projects. Marcel's expertise includes research, interdisciplinary studies, built environment ethnographies, and accounting for behavioural impacts on sustainability and wellness. Moreover, he has several publications to his credit, including, *'From Australia to AEC Industry Action,' 'The Facility Infection Risk Estimator™: A web application tool for comparing indoor risk mitigation strategies by estimating airborne transmission risk',* and *'Constructing Our Niches: The Application of Evolutionary Theory to the Architecture, Engineering, and Construction (AEC) Industry.'*

## Ethnography

Ethnography is an anthropological term that studies the methodology of understanding a group of people in their own context and how they interact with each other in a society, their power differentials, and how it affects their mutual interaction and environments. Furthermore, it understands people in a specific context through observations and interviews and being there with them without taking them out of their own context. Spending time with people in their spaces and interacting with them while they are doing their day-to-day work provides an understanding of how their existing spaces help or hinder them in meeting their needs. However, it should be noted that building design does not typically include Ethnography as part of the design process.

## Ethnography

- Systematic analyses of human interations in a defined space and time.

- Examining how the physical & social/cultural environment influences the interactions and performance of people, and vice versa.

- Consist of In-context interviews & observations

- Helps assess the relevant contextual human factors & the best proximate solution to avoid evolutionary mismatches

- Helps determine who's being impacted by what (including evolutionary mismatches) & how often

Scan the QR code to view Marcel Harmon's Linkedin Profile

# The Comfort of 3000 Years Ago to the Brain Fart of Today: What Changed?

Three thousand years ago, people intuitively understood how to stay comfortable and eventually, evolution led to the Greeks and Romans who developed technology to enhance and maintain comfort within the built environment, for example, Roman baths and radiant heating within buildings. However, after the Roman period, it was a bit like a brain fart, and progress seemed to slow down and even reverse in some geographies. Humans are subject to the forces of evolution, just like any other creature on this planet. We are part of nature, and our culture is more developed than other creatures. Our behaviour is part of what is called a phenotype, the physical expression of our genomes.

Natural selection is one of many evolutionary forces. It operates on the phenotype of an organism. If the phenotype works well for survival and reproduction, natural selection will allow it to continue, and if it does not, it is selected against over time. The built environment is part of our phenotype. It is established, but natural selection does not operate on individuals or genes. Many researchers believe that evolution operates at the group level, and things that are beneficial at the group level may not be beneficial at the individual level. Still, the stronger selection overrides it at the group level.

Taking the castles of England as an example. During the time of castles, reasons relative to social structures such as feudalism would drive building configuration and development, even if it was not thermally comfortable or defensively sound.

What proved beneficial for the feudal Lord and his group would override other characteristics. Jump forward to now, take air conditioning in the US, which is not the most dominant and effective way to maintain thermal comfort. The reason it is the dominant form is a historical accident in the late 19th and early 20th centuries. At that time, there was a need to cool down some equipment in manufacturing, and air conditioning was used to dissipate the heat. It was designed specifically for that reason, as well as humidity control.

**However, it provided some measure of cooling for people and thermal comfort. Hence, its potential was recognized and it was adapted for other uses because it was readily available. Consequently, the benefits it offered to these groups at that time and the larger society overrode the disadvantages of thermal comfort and control.**

## HUMAN EVOLUTIONARY HISTORY

- **2.5 Millions yrs:** Hunter/Gatherers as Homo specie
- **180,000 yrs:** Hunter/Gatherers as modern humans
- **10,000 yrs:** Agriculture, cities
- **150 yrs:** MEP Systems

Figure 19 Human Evolutionary History

## A Known Cycle in the Evolutionary Process

One common form of a cycle of the evolutionary process is that when new inventions are developed and made available to the masses, the people with resources gravitate to them because they are a symbol of status and recognition. However, over time they lose their elite status and become more affordable to everyone, resulting in mass adoption, and then the cycle starts again. For instance, iPhones and Tesla's are great case studies of this evolutionary process today.

Scan the QR code to follow Marcel Harmon on Twitter

# Strategies to Facilitate Optimal Resource Use at the Detriment of a Larger Group

For example, people are underutilizing gas capabilities by up to 95% for heating their home spaces, while the same gas could have been used for higher industrial purposes.

The way to tackle this trend is by looking for means to apply strategies to facilitate optimal resource use at multiple scales. The political scientist Elinor Ostrom who won the Nobel Prize in Economic Sciences, studied various cultures and the indigenous resources they were managing. She discovered eight qualities implemented by various cultural groups to help avoid the "tragedy of the commons" trap. A few of these qualities are described as follows:

- Establishing a **strong identity** among the group includes understanding and agreeing with the group's purpose and finding out their common goals and objectives.

- Having **benefits proportional to cost** ensures that benefits are not unfairly distributed among the members of a small elite group as opposed to the majority members of that group.

- Performing **consensus decision-making processes** Most people do not like being told what to do, and they like to have a say in the decision-making process. We need to be able to detect when people are cooperative or when they are defecting or being selfish. Moreover, there is also a need for graduated sanctions against poor behaviour; however, sanctions should not be draconian.

- A form of prompt **conflict resolution** perceived as fair is also required.

These qualities can be applied to a community and should be scalable to a state or nation.

### You can find more on Elinor Ostrom

About Elinor Ostrom
https://www.econlib.org/library/Enc/bios/Ostrom.html

How Elinor Ostrom solved one of life's greatest dilemmas
https://evonomics.com/tragedy-of-the-commons-elinor-ostrom/

### The Tragedy of the Commons

The concept of the tragedy of the commons came into the public consciousness in 1968 when Hardin published his essay, "The Tragedy of the Commons". The tragedy of the commons refers to the conflict for resources between individual interests and the common good i.e., society in general.

Figure 20 Large Scale Logging Reducing Rainforest Habitat

## The Potential Threat of Climate Change

Climate change continues to be an existential threat to humanity, and the US military is concerned by the potential conflicts that will arise as resources continue to be diminished and squandered. The within-group forces at this point will be stronger than the between-group, and overall, we could turn into a fractured global community. This is where Ostrom's principles come in to play. They are about figuring out ways to minimize those within-group forces, but implementing this concept globally is difficult.

## The Unusable Space in Building Design

In Marcel's opinion, the amount of unusable space built into society today is a cultural crime, and if people understood the level of waste, we would be designing buildings differently. Marcel points notes he is trying to get his designers up to speed with this waste issue during the design process.

Generally, Marcel is trying to communicate to the AEC industry about the benefits of meeting both occupant and organizational needs while minimizing energy/water use and embedded carbon. Ethnography is critical for understanding these needs and contextually meeting them. Ethnography also helps provide a more accurate assessment of the factors feeding into health and productivity impact calculations.

Overall, benefits matter. For example, the benefits from use of an under floor air distribution system (UFAD) are summarized below.
Source: *BranchPattern's National Online Conference 2021*

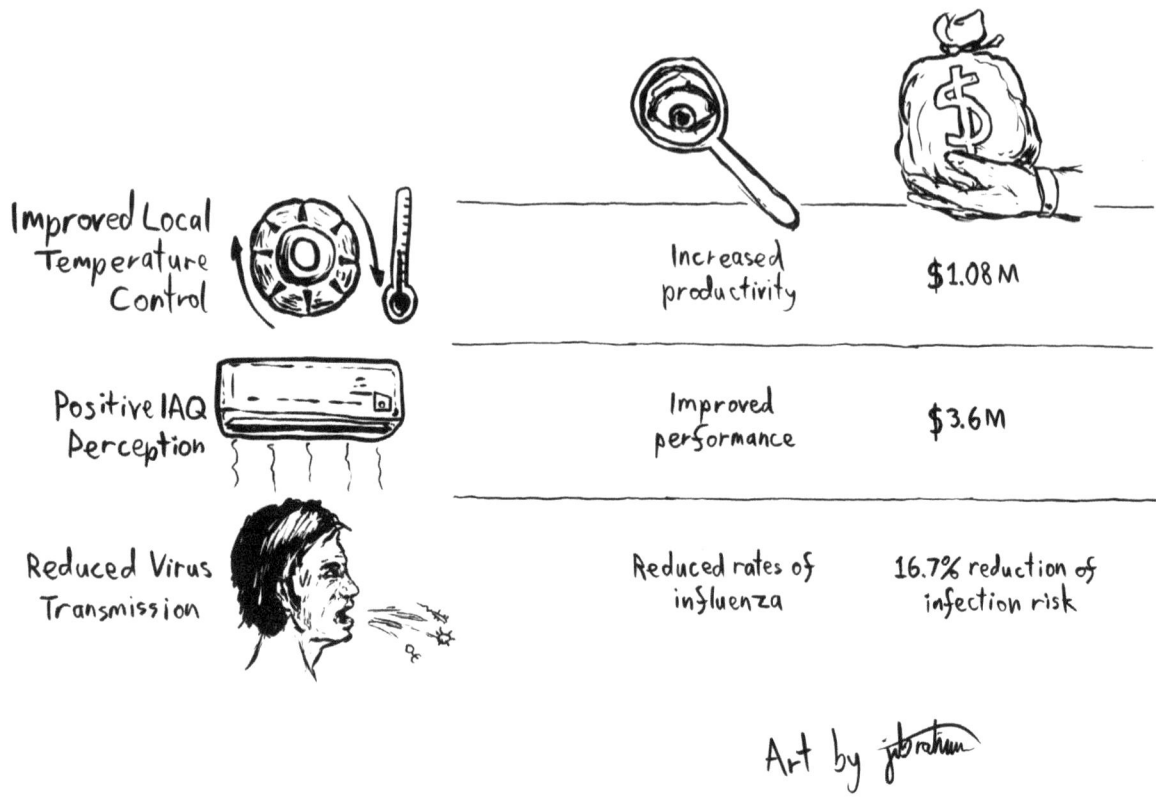

Figure 21 Benefits of UFAS Systems. Source: Marcel Harmon

## Advice to New Graduates

In terms of opting for a career path and becoming a better architectural practitioner, Marcel's advice is uniform for everyone. Each professional must develop multi-disciplinary skills or the ability to work with others outside of your immediate disciplinary area of expertise. Recognize that the AEC industry needs to better incorporate social/behavioural expertise for more successful projects and move the industry in the direction it needs to go to address climate change, environmental injustices, etc.

For instance, an architect is one part of the solution, and to deliver the finished product to the clients, it takes an assembly of crafts, trades, and disciplines. Moreover, if you want to be the focal point, make sure you are the best manager of the group the best leader from that standpoint. Therefore, get the engineers and consultants on to your projects at an early stage, consistant with the intograted design process championed by the USGBC.

Scan the QR code to visit https://branchpattern.com/

# Reader Notes: Constructing our Niches

Readers are encouraged to note their favourite insights, triggered ideas, and next steps.

# Dr. Steven Fawkes

## Finance & Energy Efficiency

Dr. Steven Fawkes understands the importance of connecting investors with the necessary expertise to deliver net-zero and regenerative infrastructure. Finance is key to the climate emergency. To quote Lucy Churchill of E P Group, "Improving energy efficiency could deliver a 45% reduction in global emissions, in a way that can be profitable". The question is, can you sell energy efficiency?

As an early mentor told me, "It is always about the money", i.e. ROI. Dr. Steven Fawkes was a great interview because he spoke about the importance of investability, project finance and investment risk. These are topics that sometimes get overlooked when architects and engineers are deep in the weeds of designing a project. The question is, can you sell energy efficiency?

*"The issue is about increasing the amount of investment that goes into energy and resource efficiency and, of course, wider sustainability. Nobody wakes up in the morning and says I want to buy some energy efficiency! "*

Dr Steven Fawkes is a member of the Investment Committee for the London Energy Efficiency Fund and a past trustee of the National Energy Foundation. A veteran within the energy and finance sector with over three decades of experience, he founded his own company, Energy Pro, in 2012 (now part of the ep group). Energy Pro was set up to accelerate the flow of capital into energy efficiency projects. Dr Fawkes is also a partner at Cameron Barney, an independent merchant bank in London, UK, providing advice and capital to clients in the infrastructure and technology sectors.

Scan the QR code to view Dr. Steven Fawkes' Linkedin Profile

# Investabilty

Dr Fawkes notes that energy supply has always been very investable, but energy efficiency is an untapped investment class. One of the reasons is that energy supply is easily measurable, whereas energy efficiency is difficult to calculate and measure objectively.

The closest thing most people understand about investments in energy efficiency is energy performance contracts. While they have been around for a while, they work primarily in the public sector and tend to be for large projects. The challenge is to make energy efficiency easy to understand, measure and invest in at scale. Investing in energy efficiency at scale ultimately has to bring in institutional and retail investors. But to achieve a big enough scale, you need to aggregate smaller projects, this is where standardization comes into play. To overcome barriers and scale-up, we need to build a jigsaw.

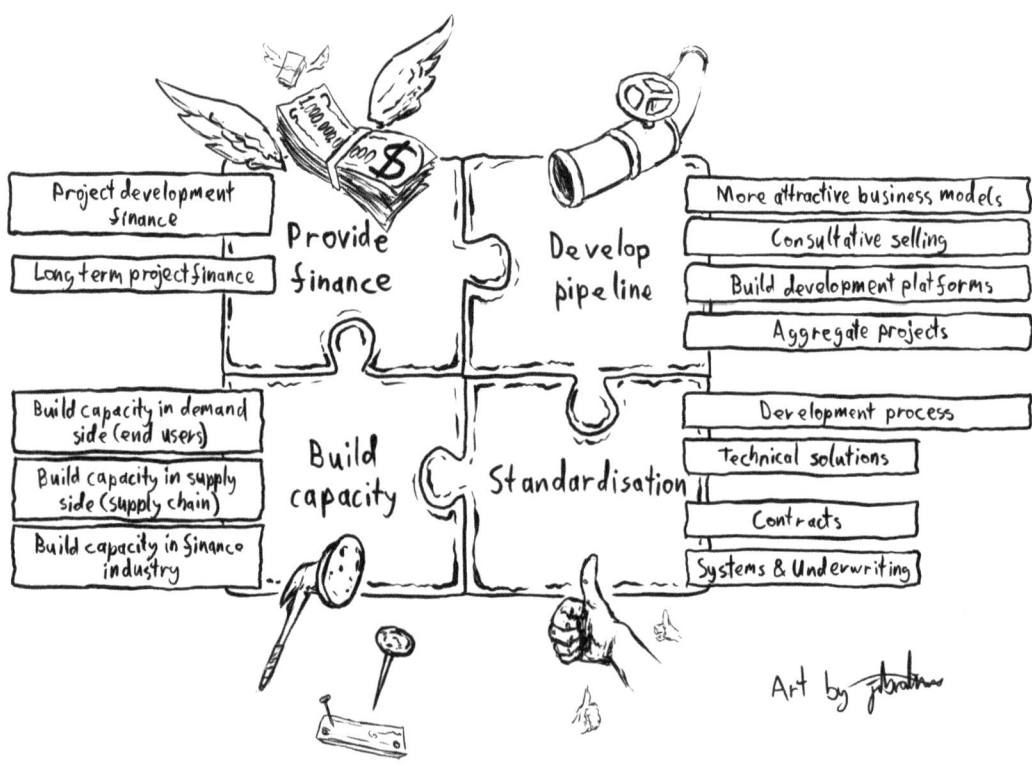

Figure 22 The Energy Efficiency Jigsaw. Source: Dr. Steve Fawkes / EnergyPro Ltd 2018

# Project Finance

One problem that has always presented difficulty in developing energy efficiency projects and solutions is the absence of generally accepted standard documentation and calculation methodologies for financing projects. It is currently difficult for investors to make sense of any project prospectus due to the numerous energy financial calculations undertaken by individual energy efficiency experts using their preferred calculation methodologies. To address this, an international project called the Investor Confidence Project (http://www.eeperformance.org/) is ongoing to bring "certified, standardized, bankable energy efficiency" projects to investment markets. The project assembles existing technical standards and practices into a transparent process that promotes efficient markets by increasing confidence in energy efficiency outcomes whilst reducing transaction costs and project performance risks. A summary of a project from development through to investment and implementation is shown below.

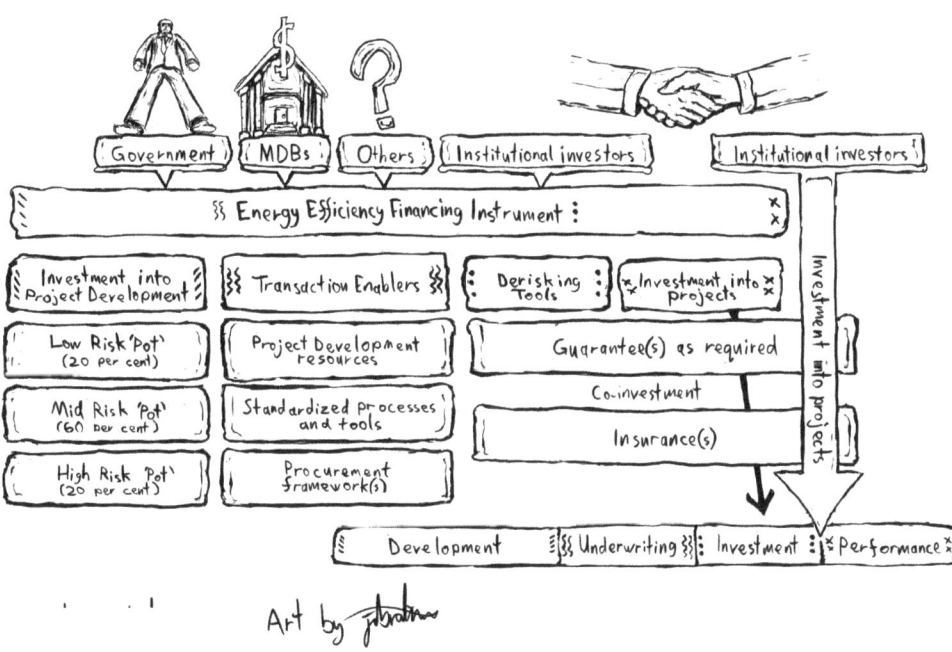

Figure 23 Energy Efficiency Project Map. Source: Dr. Steve Faeks 2021

# Investment Risk

Investment requires risk analysis; for energy efficiency, there are several questions that institutional investors must have answered before investing in projects. Typically, they include:

1. Rate of return
2. Credit risk
3. Time span
4. Measurement and recording of performance gaps between design and construction, and operation
5. Who is liable, or what redress is there for performance gaps between design and construction and operation?
6. Accuracy of energy efficiency predictions
7. Building operation / user misuse potential

Risks 1 and 2 are financial and can be hedged via insurance or performance bonds. Risks 3 through 7 are complex and require management and oversight by people with deep domain expertise. Whilst challenging, risks 3 through 7 are manageable and the Finance, Design, Build, Maintain (P3) market is leading the way in this regard.

Scan the QR code to visit onlyelevenpercent.com

# Energy Efficiency Demand

One problem is that no one wakes up and decides to buy energy efficiency. Energy efficiency might never sell if there is no awareness, information, and marketing. There is a need to create a demand for energy efficiency. The key to this lies in generating awareness of non-energy benefits such as improved air quality, less environmental degradation, and improving the asset value of the building itself. Talking about these benefits, which are more strategic than cost-saving, to decision-makers can make all the difference.

Legislation and ongoing energy policy are essential drivers of energy efficiency demand. The most critical piece of legislation has been the requirements for a minimum energy efficiency standard. This implies that when a building's energy efficiency certification is below a certain level, the owner would be unable to sell or lease the building, effectively making it a stranded asset. While there might be issues surrounding how effective enforcement might be, this is a significant market driver that should strengthen over time.

Scan the QR code to listen to Dr Steven Fawkes' Edifice Complex Podcast Episode

## Nuclear Energy

Nuclear power is the largest source of electricity in France, with a generation of 379.5 TWh, or 70.6% of the country's total electricity production of 537.7 TWh, the highest percentage in the world ([https://en.wikipedia.org/wiki/Nuclear power in France](https://en.wikipedia.org/wiki/Nuclear power in France)). Consequently, France is a low carbon economy.
Reducing carbon at scale may require adopting nuclear power generation at scale, i.e. over 50% of the country's baseload. Dr Fawkes does not entirely dismiss the possibility that the future could be nuclear in the UK and other countries. Still, he insists that no financial institution will currently buy it due to the perceived risks associated with nuclear technology.

Nuclear power at scale with improved energy efficiency from power users could be the magic bullet to reduce environmental degradation, GHG's and climate issues. Making nuclear power and energy efficiency attractive investments is one way to accelerate energy markets.

## Young Professionals

Young professionals and graduates attracted to the energy sector should look to move into the design sector, i.e. buildings, energy systems and power generation plants. This is where you can make the most impact.

Dr Fawkes believes the energy efficiency market is a trillion-dollar opportunity and feels that there is room for finance professionals, both upcoming and aspiring, to innovate and make a bigger, better market. Energy efficiency should be a fantastic investment opportunity for the next fifty years.

# Reader Notes: Finance & Energy Efficiency

Readers are encouraged to note their favourite insights, triggered ideas and next steps.

# Lloyd Alter
## The Case Against Net Zero

Lloyd Alter is one of the more interesting people in Canada's property and sustainability sectors. He understands the real issues and is not frightened to call out nonsense wherever he finds it. I particularly like his comments on the absurdities of Canada claiming to be on board with the Paris accords and sustainability yet selling F150 & RAM trucks plus "McMansions" in record numbers.

I find Lloyd's position with the case against net-zero compelling, and I have yet to see anyone push back with a good counterargument. Lloyd is a proponent of the real inconvenient truth; less is more. We need smaller and better housing, smaller and better cars and walkable communities.

*"The fundamental flaw I find with net-zero is that it doesn't say build a really efficient house. It says buy enough solar panels to compensate for the amount of energy you're using. I think that we should forget net-zero, and we should be demanding radical building efficiency."*

Lloyd Alter has worked as an architect, developer, inventor, and "prefab" promoter; he was a past president of the Architectural Conservancy of Ontario, Canada. Moreover, he teaches sustainable design at Ryerson University School of Interior Design. In 2014, Lloyd won the USGBC Leadership Award, and he currently writes for TreeHugger.com.

# Lloyd Alter Origin Story

Lloyd originally became an architect because his mother wanted him to be one, although he wanted to opt for some other fields. Once he qualified as an architect, a problem developed: actually, Lloyd did not like the buildings he designed, nor did he like driving past them. Therefore, when one of his largest clients asked for a developer, Lloyd presented himself and became a real estate developer to construct several successful condos.

At the beginning of the 2000's, Lloyd decided to continue with property development and pivoted to work with Canada's largest modular builder. Here Lloyd hired the best architects from across North America to design models that could be sold to people who valued good design. The goal was to work as an industrial designer where every practice and iteration improves the outcome. In 2002, very few people knew much about prefabrication. Hence, Lloyd started writing what became a popular blog on modular housing and became a leading authority on prefabrication, mainly because nobody else was writing on this subject. In 2004, a website called TreeHugger emerged, and Lloyd started sending them tips, which turned into a paying gig. Later on, it became a full-time writing job, and finally, Lloyd became the editor.

In addition to being an editor, Lloyd has worked as an architect, developer, inventor, and prefabrication promoter.

# The Rubble Club

In Britain, there is a community called the Rubble Club. If you are an architect, you get to join the Rubble Club if one of your buildings is demolished during your lifetime. Many historic buildings were not even finished in an architect's lifetime in the past. If there was a Canadian Rubble Club, Lloyd says he would be a charter member because the real estate market has transformed so quickly that almost every non-residential building designed by him has been demolished.

# Young People and Change

Thankfully, many young people are joining the industry. Moreover, they are not conditioned as they used to be. They do not hesitate to take decisions in challenging situations, and they have the ability to transform the industry.

When Lloyd began writing about prefabrication, the speed at which information was being disseminated, and the interest from all over North America was incredible. However, it was still a very tedious process until the high-speed internet and social media evolution. Knowledge now spreads quickly, and people are learning previously obscure things.

Today, interest in prefabrication has collided with the tiny house movement and sustainability concerns, resulting in a growing acceptance of prefabricated dwellings.

# Squarefootitis

In North America, everyone suffers from *"squarefootitis"*, which is the cost per square foot. It implies that everyone wants to pay less, and we have to transform this mindset, but it is a long, slow slog.

It was expected that there would eventually be a flight to quality. People would demand property that lasts to get the total value for their money instead of buying "cookie-cutter" disposables. However, they all want larger, cheaper per square foot houses that they can "flip." Consequently, the poor quality and externalities of this phenomenon are depressing. Hence, we all have to resist *"squarefootitis"*.

# Residential vs. Commercial Property

The most significant impact and opportunity for change is residential property because of the sheer numbers involved.

There are some excellent examples of new "sustainable" commercial property development, such as the T3 building in Minneapolis by Michael Green, but they are currently exceptions and not the norm. Commercial mass timber developments, such as the T3 building, are novelty buildings, almost a virtue-signal from the owners and occupiers. Mass timber currently has approximately the same costs as a conventional office building; therefore, in Lloyd's opinion, the T3 building is a bit like driving an early Tesla electric car. It is a signal about where the trend is moving.

Mass timber is excellent; wood sequesters carbon, and it is fashionable, but the real impact will be in residential property, making the *'Goldilocks density'* economically viable. It is referred to as *'Goldilocks density'* because it is high enough to support mass transit, restaurants, local bars, and commercial life. Still, it is not so high that we fall into anonymity and do not get to know our neighbours.

Scan the QR code to view Lloyd Alter's Linkedin Profile

# Elon Musk and His Impact

An architect named Geoffrey Warner, founder of Alchemy Architects won an award for a *"weeHouse"* he has designed and built. His success has led a director of design for Apple computers to buy one of his house designs. This is how new housing designs have started to transform trends. North American culture seems to manifest itself everywhere, and one of the most significant manifestations, in Lloyd's opinion, is Elon Musk. He told the Wired Magazine when he was asked why he was developing a dull company and building tunnels under Los Angeles; *"I hate trains, sharing with other people on the same train, there might be a serial killer behind you."* In response to Musk's assertions, Jared Walker, an urban transit expert, said that Elon knew nothing about mass transit. As a result, Elon called him an idiot, and a nerd war erupted between them.

Lloyd thinks they may both be right. However, Americans love their bubble and being isolated from others. They like their single-family homes and want to drive their car into the garage without compromising on this bubble. Moreover, they have their own theatres in their basement, and this is where society presently resides.

Elon Musk has launched his solar shingles for large single-family homes with a powerful battery on the wall and two Teslas in the garage. This innovation might be the future we all need, but these batteries do not scale and are expensive; hence, only a tiny portion of the population can access them.

However, what is the future that we actually need? Switching cars from gasoline solves part of the problem, but it does not solve congestion, nor does it solve parking or the whole urban sprawl issue. According to Elon, autonomous (self-driving) cars are a way of increasing urban sprawl, enabling us to enjoy a martini in our car or watch a movie as our car drives us to our home instead of concentrating on the road.

Figure 24 Elon Musk

# Removing the Duck Effect with Battery Storage Technology

The duck effect is related to what is known as the 'duck curve'. This phenomenon occurs particularly in climates like California and Arizona, where abundant sunny days and solar panels produce a lot of energy during the daytime, and whilst electricity demands peak in the early evenings.

Solar energy production in climates like California and Arizona during the daytime is more than demand. Therefore, power utilities are usually shut down and working below optimum capacity. However, as people arrive at their homes in the evenings, there is a sudden huge demand, and solar panels cannot generate the required power, and demand from the power utility surges. Subsequently, this demand tappers off as people go to bed. This process is called the 'duck curve'. There are two different points on the duck curve, which are shaped like a duck.

These points are prominent showing energy demands during the early morning and the evening when the utilities face a problem meeting demands.

Scan the QR code to listen to Lloyd Alter's Edifice Complex Episode

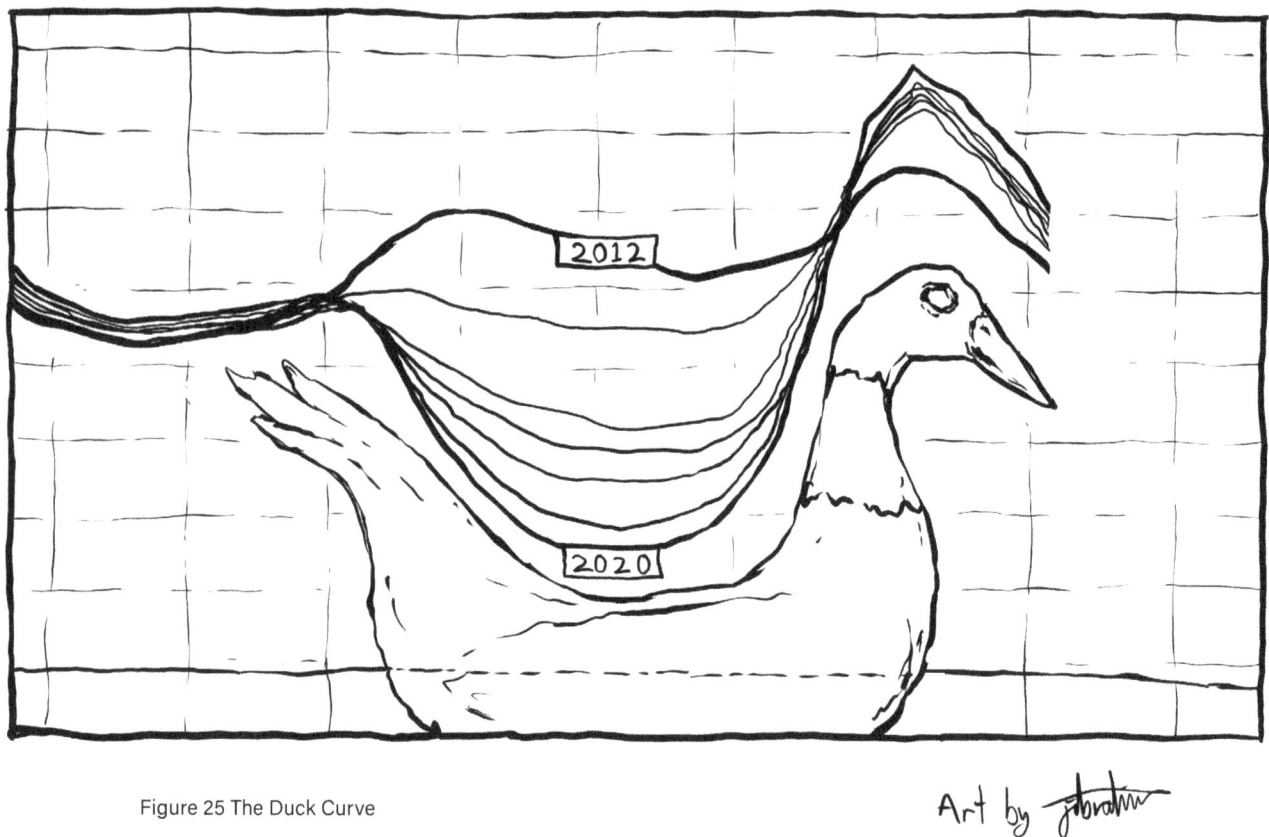

Figure 25 The Duck Curve

In Australia, power distributors had a problem, they were experiencing brownouts at 6 pm because they could not meet demand. The solution was provided by Elon Musk, investing $50 million into a giant plant and filling it with batteries; basically, "killing the duck". The electricity from the solar panels is stored in the batteries during peak production hours. Between 6 pm and 7 pm, the battery packs can provide the power needed for 30,000 houses during peak demand. Lloyd believes this efficient solution will be deployed in other countries as solar power becomes more ubiquitous.

# The Case Against Net Zero

Lloyd objects to a Net-Zero scenario. Numerous people are building houses and boast of being Net-Zero by generating enough power in the summer to repurchase it in the winter. It implies that there is a whole electrical system built, designed and maintained with generating plants to support demand during the months of December, January and February. We are not paying for power in July, August, and September, and now have the money to put the solar panels on our rooftops. However, what about the people who have rented an apartment or the people living in structures without any rooftops?

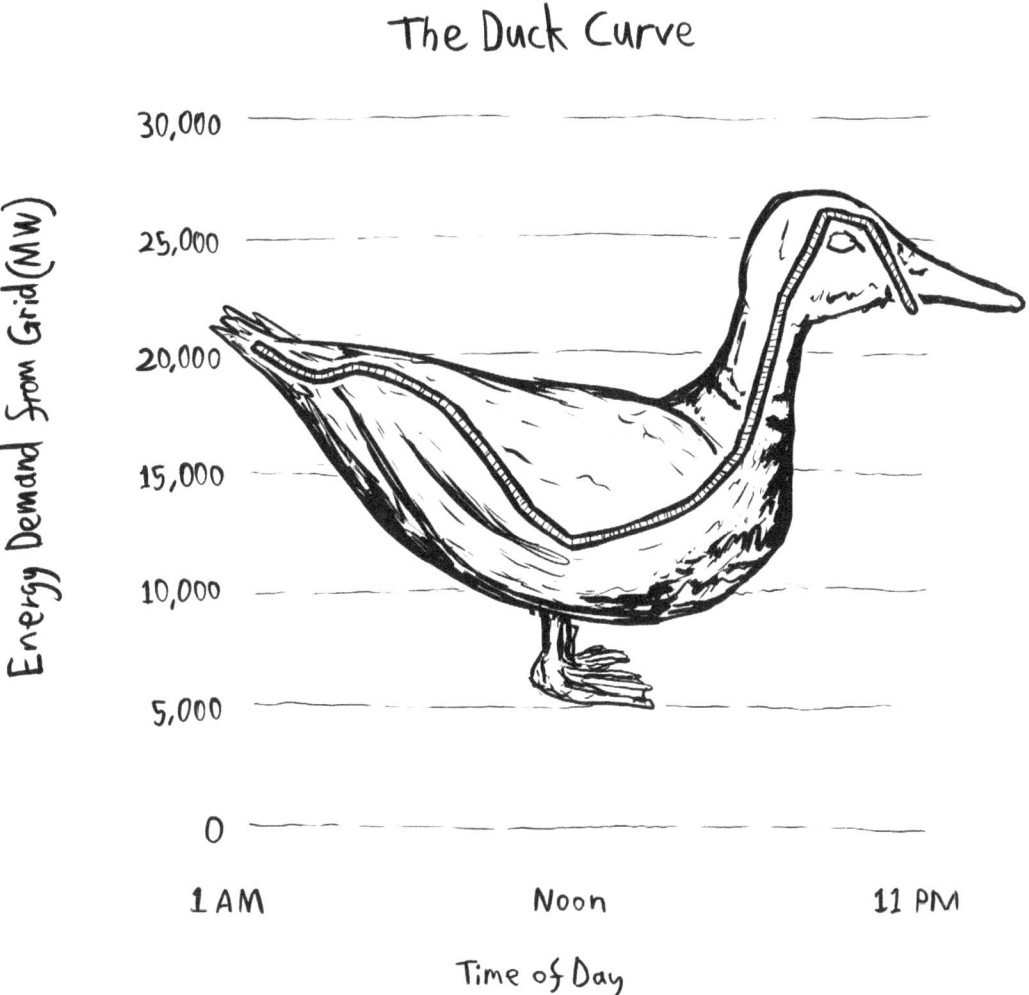

Figure 26 The Duck Curve

Moreover, what about those who cannot take advantage of Net-Zero? The fundamental flaw with Net Zero is that it does not allow us to build an efficient house. It says buy enough solar panels to compensate for the energy you are using. We should forget about Net-Zero and demand more radical building efficiency, and we should be building houses that do not need much power at all, buildings that you can run with a couple of 100 watts, and then we can afford to maintain a power grid. Hence, the further we get into Net Zero, the more trouble we get into. The Net-Zero system does not mean anything for comfort, and it does not address resilience. What happens when systems break down? What is the backup? "These are the things that we have to worry about, comfort and resilience, not just the cost of energy, which is not a factor that people care about anyway."

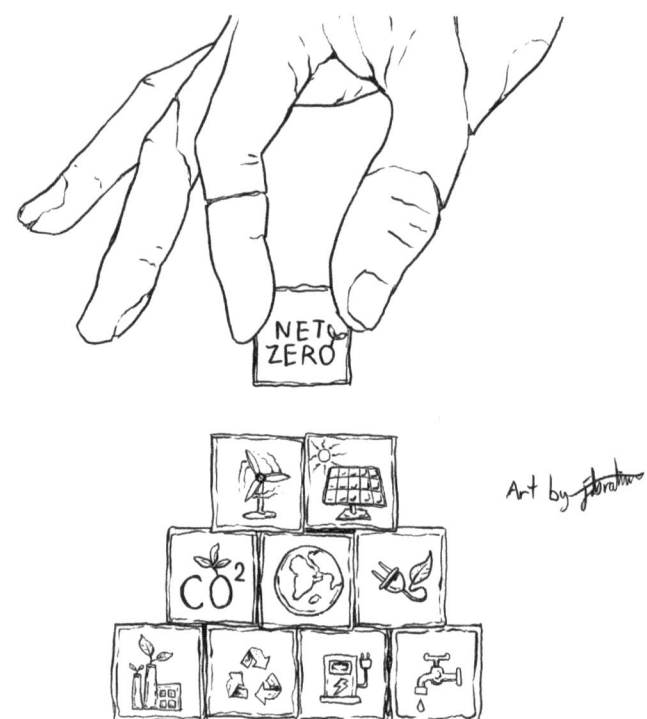

Figure 27 Sustainability Heirarchy

## How to Change Things

We in North America need to mimic the systems developed in Europe, as they are setting high standards for their building codes. In Belgium, you cannot build anything that is not up to a high standard, and they are tightening these standards every time there is a revision to building codes. Some people say they cannot follow all the standards and codes due to the cost of housing, but you can build a little smaller and a little more efficient with higher performance. When you invest in your house, you can get 1,500 square feet instead of 2,200 square feet, but with care and thought, it is going to be a good 1,500 square feet. Less is more!

Scan the QR code to visit Lloyd Alter's website

# Reader Notes: The Case Against Net-Zero

Readers are encouraged to note their favourite insights, triggered ideas, and next steps

# Henry Gordon-Smith

Agritecture

Henry lives at the intersection of the built environment, agriculture, high technology and sustainability. Henry's business, Agritecture Consulting, pulls these four areas together to address one of the societies pressing questions, *"how do you feed and sustain high-density populations?"* Also, *"how do you do this not just at a minimal survival level, but at a high level with highly nutritious and healthy produce?"*

The mission Henry and his team are on is not easy. However, it is probably vital to societies future success as human populations flock to high density living in "mega cities" around the world.

I find Henry inspirational. He is living his core values and realizing his mission on projects worldwide. He is, making a difference, plus he is trying to do it at scale. In my opinion, he is one of the most interesting people in the property industry right now.

***"When we think about the future of cities and how we are going to unlock sustainable urban development, it's my philosophy that agriculture will be the key to unlocking sustainable urban development."***

Scan the QR code to view Henry Gordon-Smith's Linkedin Profile

Henry Gordon-Smith started Agritecture.com in 2011, a leading media platform covering the news, business, and design of how agriculture can integrate with the built environment. Subsequently, he co-founded the Association for Vertical Farming (AVF) in 2013, which grew to have approximately 50 business members, including several multinational companies. The AVF has since hosted its annual Summit in Beijing (2015), Amsterdam (2016), and Washington DC (2017). In 2017, Henry resigned from his role at the AVF.

In 2014, Henry started Agritecture Consulting, the premier urban agriculture consultancy assisting a global portfolio of projects. Agritecture Consulting is an extensive company, and it has provided expert guidance to over 150 clients in 35 countries, including entrepreneurs, multinational companies, architecture firms, municipalities, and educational institutions. Whatever project Henry is working on, his primary motivations are the triple-bottom-line: people, planet, and profit.

# Vertical Farming

Figure 28 Sky Vegetables Top Greenhouse, New York City. Source: Agritecture Consulting 2021.

Vertical farming is a high-tech type of urban agriculture. Suppose you imagine a spectrum of urban agriculture on the low technology side. In that case, there are traditional community gardens and other soil-based farms within the vacant spaces of a city. As multiple things become more advanced, we can use engineering to integrate urban farming into rooftops and facades, but these options are still soil-based. As we acquire better technology, we can control the growing environment, creating a structure for year-round growth. As a result, we then attain 'controlled environment agriculture', typically known as *greenhouses*.

In the next iteration, there are the containers, a unit that has been retrofitted to have vertical farming systems within. Vertical farms are three-dimensional farms, and they typically use hydroponic systems; these systems are stacked over each other to maximize the vertical space, thereby reducing footprint. In vertical farms, the climate and PH can be controlled, providing the environment required by a plant to thrive no matter the season.

Vertical farms come in different sizes. For instance, a basement can be converted into a small vertical farm with 50 square meters. It can also be a container, or a big warehouse to provide scale. Thus, vertical farming has the advantage of a small footprint with potential high-volume farming.

## Dynamic Environments Required

Throughout their life cycle, plants change, and these changes depend on the type of plant grown. Some farms will grow cannabis plants and some will grow tomatoes; this is a significant shift in the nutrients and light we would want to provide throughout its growth cycle. Moreover, there are also various harvesting strategies for each type of plant. For example, a tomato will not start producing until several months after we plant it, while a lettuce plant will be ready to harvest after only 30 days, sometimes less if you grow it on an indoor farm.

What Agritecture does is analyze all these subtle differences and store them in databases to help clients plan how they can utilize their space and identify the operational challenges. In some cases, there are minimums and maximums. For example, it is unlikely that a 50 square meters farm will be profitable, and there is a need for a scale for those plants with longer life cycles. For instance, some plants require more labour costs, and the heating, ventilation, and cooling system for other plants remains a challenge, because as plants grow through their different stages they create microclimates.

# Unlocking sustainable urban development

Vertical farming and greenhouse agriculture in the city are the keys to unlocking sustainable urban development. The things we are now learning about how biological systems are treated and the different engineering components help us comprehend the circular economy and enable us to understand how various resources communicate between buildings in a way that we have never experienced before. There is no other emerging and sustainable technology which can embrace the nexus of food, water, and energy waste better than urban agriculture. Therefore, it is attractive for cities to get behind these ideas and for developers to explore investment opportunities.

# Retrofitting

Retrofitting can be a big challenge. In the case of say a 10,000 square foot farming facility, the HVAC requirements are extreme, and retrofitting this kind of demand can be cost-prohibitive. However, in some markets, the conversion of retail into smaller micro farms is feasible and can attract foot traffic leading to the revitalization of an area.

**Price of food, food safety, and the shift to vertical farming**

In the future, as food prices rise, the incentive to move to a vertical or urban situation will be enhanced because the current mass production and relatively cheap cost of food are not sustainable in the long run. In addition, externalities in the current food system are not being accounted for when we pay for our food products; one of these is food safety.

In 2019, the United States CDC announced on several occasions that we should not buy lettuce from stores due to E. coli outbreaks. Food Safety is another justification for growing food indoors because we can create clean growing environments. There are also externalities from mass storage and transportation and storage of food products, such as energy costs, water overuse contamination, food waste and misallocation of capital plus human resources. According to a CDC statement, *"Greenhouse-grown and indoor farm-grown romaine are safe to eat, but outdoor-grown romaine should not be eaten. If it's in your fridge, throw it out now."*

Although indoor farm produce is currently more expensive than outdoor farm produce, it is getting cheaper. The externalities noted above and their increasing costs will over time be the dramatic transformation that will justify investments in urban and vertical farming.

Figure 29 Urban Farming Using a Container

# The Situation in Northern Canada and the Way Forward

Considering obesity and heart disease rates in the Northern communities, there have been real consequences from the lack of fresh food. It should be noted that before the modern era, indigenous communities were able to grow their own food and sustain themselves. Now they have become dependent upon imported processed food.

The climate in Northern Canada makes it very hard and expensive to obtain fresh produce. Due to difficult logistics, high imported food costs provide incentives for indoor farming in this region at the individual and market levels. In addition, improvements in vertical and micro indoor farming coupled with advances in decentralized energy systems could jump-start the circular economy for these communities. The bottom line is that more local fresh produce could provide jobs plus improve health and well-being for Northern communities.

# Water and Agriculture

Figure 30 Melia Desert Palm, UAE. Source: Agritecture Consulting 2021

It is estimated that about 90% of the Canadian population lives within 100 miles of the US border, and much of Canada's water simply flows away from the population towards hard-to-reach regions. This means that delivering water to population centres is expensive, despite Canada having abundant freshwater resources. As freshwater scarcity grows in the US, it will demand water from Canada, which may bring interesting geopolitical issues. Vertical farms provide controlled environments that can minimize the use of pesticides and water via demand-driven drip irrigation. Moreover, urban vertical farms also have the advantage of being located in regions where the infrastructure exists to deliver water at scale. Indoor farming systems save at least 70% of water relative to conventional farming and sometimes even up to 90%. As water costs become a significant factor in food production, vertical farms offer opportunities to save water at scale.

Freshwater scarcity is growing in the US due to industrial farming, water table depletion, pesticide runoff, and industrial pollution. In addition, "virtual water," that is, water moved across borders in the form of food products high in water consumption, is creating water shortages in "water exporting" regions and leading to the development of the unsustainable regions that *"import virtual water."* In the long term, water scarcity will lead to high food prices.

Scan the QR code to listen to Henry Gordon-Smith's Edifice Complex Episode

Figure 31 Agritecture Design Support. Source: Agritecture Consulting 2021

## Investment Potential in Agritecture

Henry has been blogging about urban agriculture since 2011, when there were no commercial vertical farms in North America. In 2020, Henry identified 350 active vertical farming companies. Increasingly, "Ag. Tech" is an investment theme attracting entrepreneurs that want to influence sustainability and agriculture. However, there is still a long way to go. A lot of these "Ag. Tech" startups do not consider the farmer properly.

Farmers have to adapt, and startups need to learn about each other with a merger of farmers with their experience and "Ag. Tech" with their technical expertise, urban vertical farms would make a productive combination for the future.

The Silicon Valley attitude applied to "Ag. Tech" simply does not work, because in terms of capital, a long-term view is required. It would be good to see big engineering companies investing in vertical farms, and cities finding ways to create their own farms to provide as a resiliency strategy for population centers.

# A Tool That Can Help You

Henry and his team aim to help thousands of people globally understand the methods necessary to grow food in the city and get their ideas onto paper, whether entrepreneurs or investors. Therefore, they have built an online planning software tool on their website (URL: https://design.agritecture.com). It is a free concept tool where you can enter any relevant information, such as your location, the type of farm you want to build, and the kind of crop you want to grow. As a result, you will get a recommendation of already-existing similar farms and a sample project plan on what it typically takes to build out a farm.

Additionally, there is also a paid subscription level, where you can build a 10-year projection of a farm and an unlimited number of projects. This level is aimed at real estate developers, entrepreneurs, architects, and cities to assist their feasibility studies.

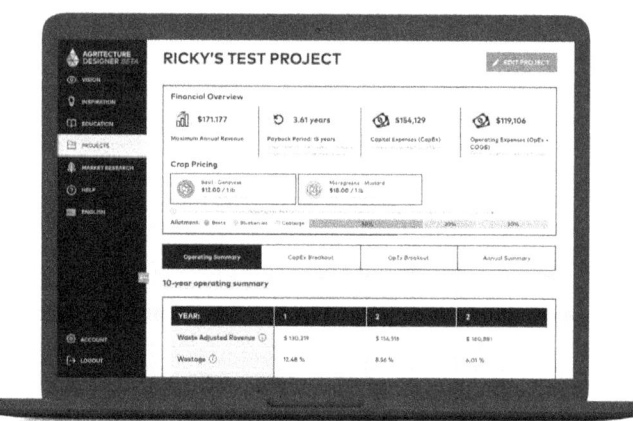

Figure 32 Test Project Dashboard. Source: Agritecture Consulting 2021

Scan the QR code to visit Henry Gordon-Smith's website

# Reader Notes: Agritecture

Readers are encouraged to note their favourite insights, triggered ideas, and next steps.

# Dan Nall

Design Principles & Discipline

Dan's career arc has taken him from early energy modelling in the 1970s to engineering, architectural and LEED fellow today. Whilst, in my opinion, Dan is first and foremost an engineer, his architecture and sustainability skill sets provide him with a unique perspective when designing building systems. Dan is that rare thing in building engineering; he is truly a systems thinker.

During Dan's podcast interview, I could not help but think he was the best teacher I never had. In my opinion, Dan is precisely the right person to teach building systems engineering and get people excited about careers within the built environment. Dan's episode of the podcast is a master class on being a building services design engineer.

*"Design your projects based upon a set of desired outcomes, which could include diversified demands, such as save the planet, come in under budget, and make the client(s) happy etc."*

Dan Nall has been engaged in various energy efficiency projects in buildings since 1977, starting with his graduate work in computer simulation to optimise the thermal performance of buildings. In addition, he also worked in various roles during his diverse career as a building science researcher, building energy consultant, mechanical design engineer for buildings, and sustainability consultant.

A graduate of Princeton and Cornell universities, Dan is a registered architect and a professional engineer. Moreover, he is an ASHRAE life fellow, a fellow of the AIA, and a LEED fellow. He has held senior VP positions with Syska Hennessy and WSP Flack + Kurtz during his remarkable career.

Scan the QR code to view Dan Nall's Linkedin Profile

Some of the noteworthy projects in Dan's resume include BASF US Headquarters, Newseum in Washington, DC, Hearst Corporate Headquarters, US Embassies in Sofia, Bulgaria and Cape Town, South Africa, Clinton Presidential Library, New York Times Headquarters and Alcoa Corporate Headquarters.

## Origin Story

Dan Nall's initial aspirations to become an astronaut and physics major could not materialise due to a medical discharge from the US Navy owing to an injury. Nevertheless, after completing a degree in English literature, Dan had to figure out his professional career. He realised that he enjoys solving real-world problems, and this ability led him to pursue a career in architecture and engineering.

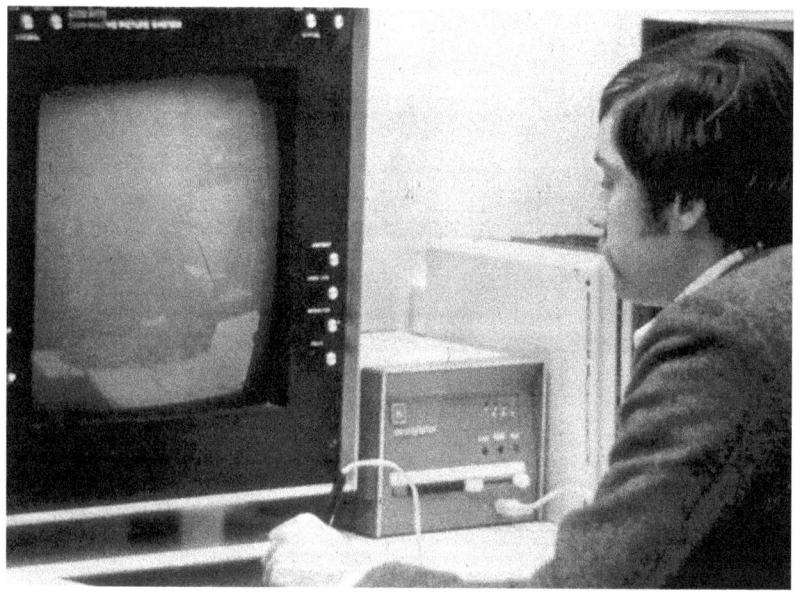

Figure 33 Dan, in the 1970s, using "state of the art" computing technology. Source: Dan Nall 2022

# The Old-Fashioned Methodology Works

*"In this business, there are too many individuals who become fascinated with the newest technologies, and they become bound and determined to implement a particular design."*

People tend to incorporate the latest trends in their building designs rather than sitting and contemplating the actual problem. At Cornell University, Dan studied a subject titled *'the discipline of design'*, which necessitated performing research and establishing a context to work i.e., a real-world project. Subsequently, he would be required to develop a list of desired outcomes. The next step was to create the hypothesis for the problem(s) to be solved and produce the desired results.

Figure 34 Torre Banco Macro Building, Buenos Aires, Argentina. Source: Dan Nall 2022

Next, the details were added by putting layers of yellow trace paper on the original drawing, along with some additional detail, and looking at the design for the pros and cons. Finally, a decision was made to add the detail or pull it off. In this manner, he learned a framework for testing every level of detail to see if it works or not.

From this Cornell University course, Dan learned that design is an iterative process, different from the numeric, quantitative way of solving problems employed by most engineers.

## Performative Sustainability vs Fundamentals

Control sequences are fundamental. The idea of structured programming and creating testable subroutines is the basis of sound systems design. Each huge task contains multiple subroutines and sub-tasks. As a building system designer, you should review all the system inputs to ensure the desired outputs. Therefore, a capable systems engineer will never forget that, *"When designing a concept for a system... design the control sequences first".*

Take, for example, a strategy implemented in a building that requires a lot of heating, such as NYC. There is no such thing as free cooling. Free cooling is the process of throwing away the heat that could be used for heating your domestic water systems. To achieve a heat recovery strategy, we have to look at the various configurations that the building will find itself in and the responses to deal with these circumstances like the conjunction of occupancy, internal heat gain and required ventilation. It is important to determine how the system will react to these circumstances. Finally, based on an in-depth investigation of the design space, we can select the required equipment and the most optimum system configuration.

# NYC Local Carbon Emission Law

Reducing the carbon footprint of a building does not come cheap and effective products are rarely available. Hence, low energy solutions are required, which will need the building design engineers to deliberate differently and maximise the recycling of heat gains to minimise the heat loss.

NYC local law 97 is actually a welcome development. Its main objective is to restrict the amount of carbon emissions allowed for a specific building type, leading to many existing and new building owners scrambling to comply with the law and avoid being fined. The only source of disappointment concerning NYC local law 97 is the taxonomy of buildings. The law is based on International Building Code (IBC) assembly occupancy groups focused on fire hazards rather than the EPA and Energy Star building types oriented around having a relatively similar energy performance.

This means that there will be different energy consumption levels, and when you put a metric on the various occupancy groups, someone will get a free pass, and another's will not.

## The Separation Strategy

One strategy I like, which improves energy efficiency and persistence of comfort, is to separate ventilation and dehumidification from sensible heating and cooling. For example, a dedicated outside air system with radiant heating and cooling for sensible loads can be a game-changer for user comfort and energy efficiency.

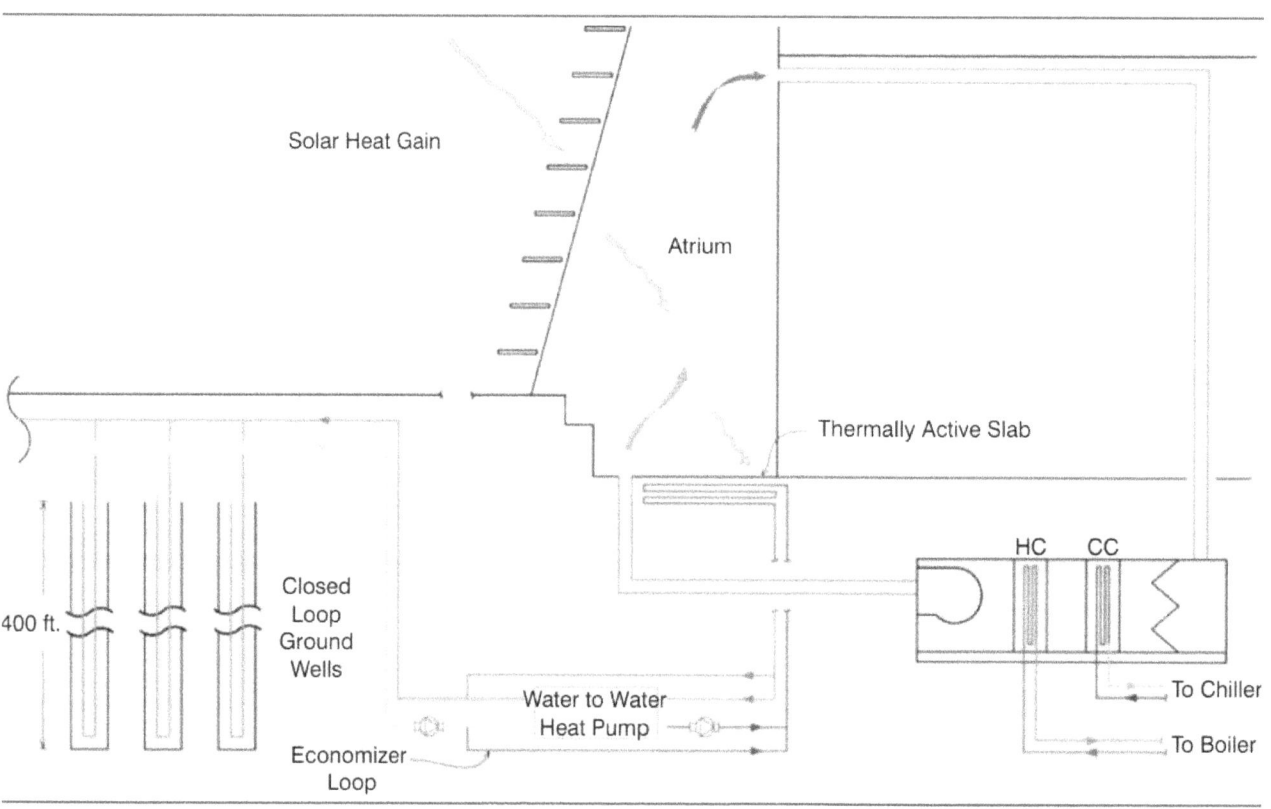

Figure 35 Heat Pump System with Thermally Active Slab. Source: Dan Nall 2022

# HVAC Design and Energy Transportation

Few people understand that a significant component of building energy consumption is the amount it takes to move the energy around. Heating and cooling are not consumed at the point where it is created; you need energy to move it from the point of creation to the point of use. In mathematical terms, moving one unit of heat or cooling from the point of creation to end-use with air systems will take about four to ten times more energy compared to using water systems. Therefore on this basis, radiant heating and cooling systems are preferred design solutions.

# Final Thoughts: Outcome-Based Design

Building engineering designs should be based on a set of desired outcomes; these desired outcomes can range from saving the planet to making a client happy to meet the budget. During the course of the building systems design process, you should be aware that you will have to give up some of the desired outcomes, especially when it is not within budget. When going through the design process, you become aware of what you can achieve and what you cannot and what you can save for the next project when the technology is better or with a larger budget client. If architecture and engineering are pursued in this fashion, i.e. partially pragmatic and partially aspirational, our built environment will continue to improve.

# Advice to New Graduates

For fresh graduates with mechanical engineering degrees exploring job opportunities, it is imperative to obtain a broad knowledge and experience base early in their professional careers.

*"...the mixture of keeping first principles in mind and at the same time, getting in the field to see projects getting built, and seeing it not work out, seeing what can go wrong and how you make it work".*

The broader your base, the better engineer you will be; never lose sight of first principles. Moreover, you must keep developing spreadsheets and the controls sequences for your analysis. For example, you can explore questions like these: *'what is the required dew point temperature for this ventilation air stream to maintain 74 degrees Fahrenheit and 50% relative humidity on the top floor of a building which is 1200 feet tall'.* You can first draw out the building and its room outlines for a possible solution. Subsequently, you can carry out an analysis to help you understand what is going on from a first principle's standpoint. Your sound fundamentals will always take you through to a plausible design solution.

Scan the QR code to listen to Dan Nall's Esifice Complex Episode

# Reader Notes: Design Principles & Discipline

Readers are encouraged to note their favourite insights, triggered ideas, and next steps.

# Glen Spry

## The Coming Energy Pivot

Glen is an agent for change in an industry that resists change. He strongly believes that the application of technology, IoT and decentralisation can get us out of our impending energy crisis. As C.E.O. of SensorSuite Inc, Glen puts his theories into practice by applying IoT technology and software solutions to existing and new buildings.

Given the conservative nature of the energy industry, I see Glen as a great example of leadership using entrepreneurial zeal to tackle significant scary problems. Glen and SensorSuite Inc prove that small and medium-sized firms can make a difference and take a leadership position in established industries.

*"The property industry right now is filled with soloists... but we need an orchestra. We need all these little pieces to come together to really provide that killer value proposition. We've got to come together to create that symphony... partnerships are fundamental to the transition."*

Glen Spry has over two decades of experience in business development, sales management, and energy market strategy. As C.E.O. of sensorsuite, an Internet of Things (IoT) company, Glen focuses on optimisation and efficient HVAC systems in buildings for real estate investment trusts and other portfolio managers. Furthermore, Glen is a skilled business manager within the energy sector.

# Existing Energy Infrastructure

One fundamental issue is that most energy grids, particularly in Europe, are based on old consumption and technology models, and they are a bit like *"model T Fords"* in a world that needs a *Tesla*. Existing energy infrastructure must be modernised, and it should incorporate demand management software and machine learning. In addition, the mix of energy sources, *e.g.* gas, wind, solar, battery, nuclear, needs to be widened to offer better demand response and resilience. We are at a turning point in infrastructure history where buildings and transportation are electrifyingly creating a step demand curve. Still, existing infrastructure is nowhere near ready for this step-change in demand.

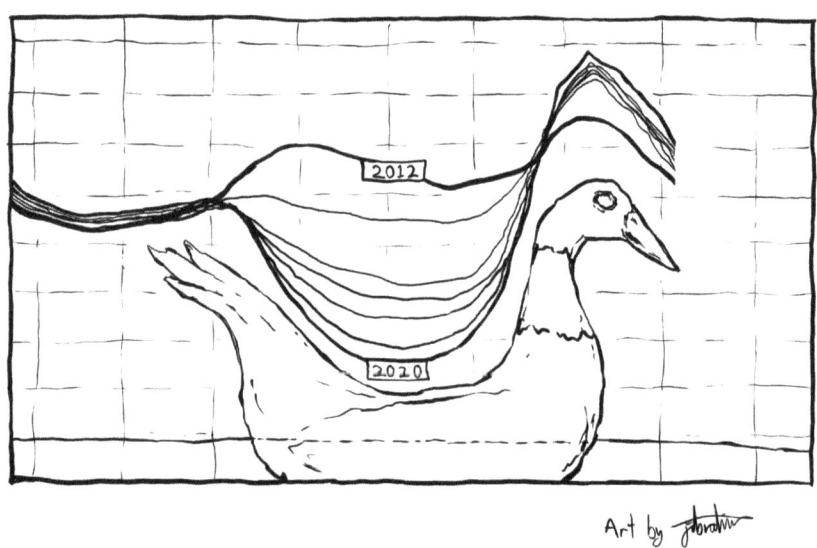

Figure 36 The Duck Curve

Scan the QR code to view Glen Spry's Linkedin Profile

## D.O.U.G.'s

Another reason the energy market is behind the development curve is D.O.U.G.'s (Dumb, Old, Utility, Guys). These older people have been in a unionised environment and have a fixed worldview with their eyes focused on their retirement. They enjoy the status quo and do not want to change the processes and methods necessary to modernise their businesses. They are *"ticket punchers"* doing the bare minimum to avoid being fired. Clearly, this is a caricature, but as with many things said in jest, a large element of truth makes it recognisable.

In summary, the current situation for many Western countries is that ageing energy infrastructure and dependence on fossil fuels are all driven by D.O.U.G.'s. This does not mean that change is impossible. However, it does make change political and difficult. Humans are short-term thinking creatures, and nobody thinks of the electricity supply until it is not there.

Scan the QR code to listen to Glen Spry's Edifice Complex Episode

# Consumer-Driven Energy Transitions

Consumer demands will drive the coming energy infrastructure transition, and I currently see a consumer-driven paradigm change at the utility level.

Consumers are beginning to utilise alternative sources of energy, e.g. solar and wind. They do not want to rely on an archaic energy system with expensive, unreliable, and dirty energy sources. Consequently, this situation leads to demand and supply issues for utility companies. For example, in California, U.S.A., where there has been a significant uptake in residential solar energy, a phenomenon called the *"duck curve"* occurs. It is low gross energy demand during the day when the sun is up and a massive surge in demand in the evening when solar power generation drops precipitously. As a result, most utility companies struggle to deal with this big swing in demand. In addition, the drop in revenue to utility firms from low daytime energy use feeds into a cycle of reduced investment in the energy grid leading to ever more unreliable and extended use of ageing energy infrastructure.

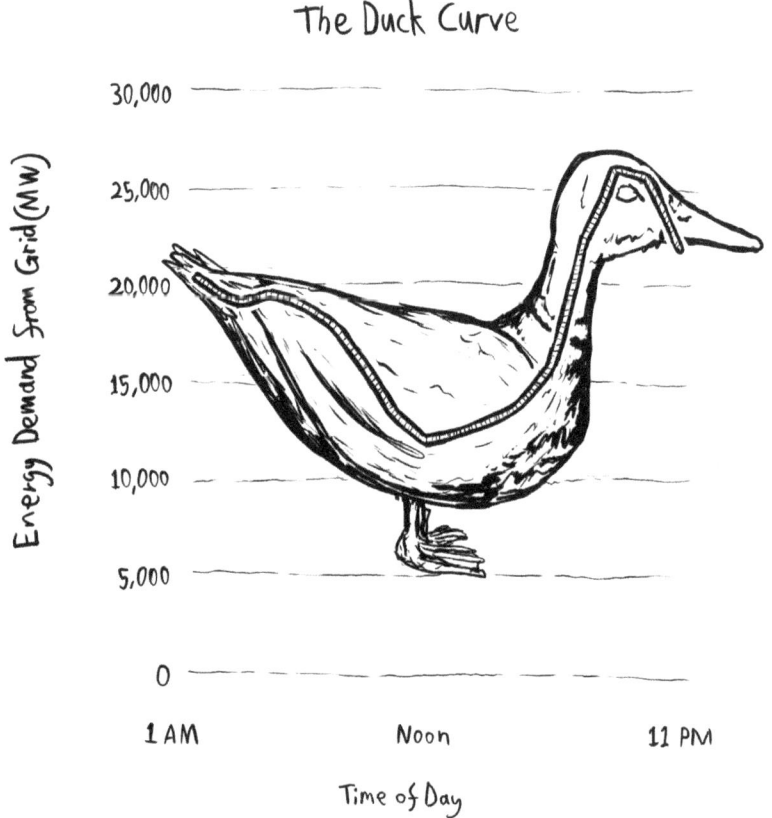

Figure 37 The Duck Curve

Glen believes that there is more than enough maturing technology to modernise the energy infrastructure, and costs are falling to the point where it makes sense to spend CAPEX and start upgrading energy grids. The D.O.U.G.'s obstructions have a limited time period, as they will eventually age out of the industry and retire. The main obstacles going forwards are likely to be political. Governments are good at reacting to the immediate crisis but terrible at advanced planning with no immediate political gains.

Figure 38 Even Relatively New Assests Can Become Stranded. Source: SensorSuite Inc 2021

# Stranded Assets

As alternative energy sources are being adopted (e.g., wind, solar, and battery) coupled with various energy reduction strategies (e.g., NetZero construction and retrofits, demand management via software), the possibility of "Stranded Energy Assets" grows.

Fast adoption of renewable energy sources at the consumer level can turn the old energy grids into unresponsive "stranded energy assets" within the next five years. Even relatively new assets can become stranded. Stranded assets are "assets that have suffered from unanticipated or premature write-downs, devaluations or conversion to liabilities". For people wanting to move to a renewable energy future, this would be seen as "creative destruction." For D.O.U.G.'s and politicians, this is a disaster to handle and resist.

Governmental resistance to "Stranded Energy Assets" will likely be prevention via legislation preventing consumers from moving off the existing energy grid. I believe that it will become illegal for the consumers to be "off-grid" for their "own safety," and a connection fee (Latin for tax) will be applied to allow a funded, slow transition to update the existing energy infrastructure and pay the D.O.U.G.'s pensions.

Figure 39 Managing Comfort at the Individual Room Level. Source: SensorSuite Inc 2021

# Transition

We are told we are in the age of gas but rapidly approaching the lithium and energy storage age. However, current upstream gas and coal energy production and transmission waste approximately 54% of the energy produced, and efficiency in production and consumption go hand in hand. Glen believes real-time demand management, local energy production, and storage plus micro energy grids are the future.

Smaller energy grids with local battery storage to manage peaks and troughs in energy demand are more complex than large energy grids with economies of scale. The issue is that large-scale energy girds require stable demand curves, which are becoming harder to acquire. The small and local micro-energy grids and storage primarily provide resilience and faster demand responses. Nonetheless, advancements in data transmission and software technologies can manage the complexity of "more grids".

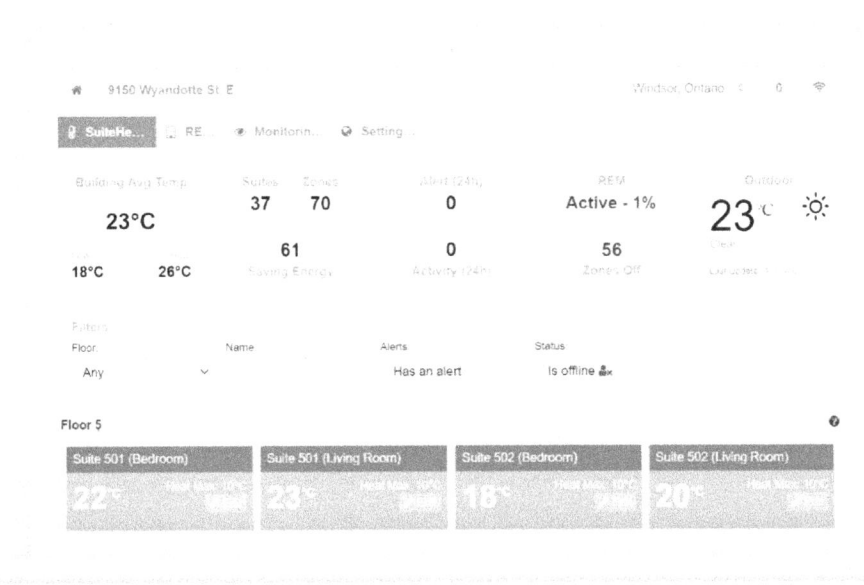

Figure 40 Managing Comfort at the Individual Room Level. Source: SensorSuite Inc 2021

## Energy Orchestra

Partnerships are fundamental to any transition. As Glen puts it, "what is needed is more like an orchestra than soloists." Unfortunately, the industry is currently a hot mess of soloists rather than partnerships. By the end of the decade, what is needed is multiple acquisitions and roll-ups, giving room to the companies with a vision of the future.

## Industry Opportunities

We are entering a period of necessary change when technology provides multiple options to solve many older problems. This "fourth industrial revolution" era will leave its mark on energy infrastructure renewal.

Glen's view is that demand for good people will outstrip supply in the future. Therefore, as long as you know your stuff and have something to offer, it really does not matter what gender you are or your background, and what will really matter is what you can actually deliver. The energy industry provides fantastic opportunities for employment and rewarding careers.

Scan the QR code to visit sensorsuite.com

# Reader Notes: The Coming Energy Pivot

Readers are encouraged to note their favourite insights, triggered ideas, and next steps.

# Dr. Rochelle Ade

Green Building Certification is Not Working

Dr Rochelle Ade is a rebel, a very well educated rebel, which is why the property establishment finds it hard to cancel her. Rochelle's journey has taken her from green building advocate to concluding green building certification systems should all be abolished. In effect, Rochelle says the "Emperor has no clothes", and it is about time we all acknowledged this.

None of Rochelle's conclusions are based on opinion. They are based on years of experience and detailed research during her doctoral studies. The lack of evidence that green buildings and green building rating tools delivered buildings that would help save the planet was troubling and could not be ignored. Someone had to call this out, and Rochelle, to her credit, has.

As soon as Rochelle called out green building certifications as "greenwashing" in our interview Robert and I were hooked. Rochelle is a leader and agitator for real change.

*"Reality is we build something, fit it out, and then people live and work in them. There is thus a symbiosis between the building and the occupants... but every single rating tool I've ever come across divorces the two."*

Dr. Rochelle Ade has worked in construction in New Zealand and globally, primarily as a project manager and green building consultant. Rochelle's interest in green building was sparked during the formation of the New Zealand Green Building Council in 2007. Subsequently, she carried her passion to the Middle East. She was involved during the establishment phase of the Emirates Green Building Council while heading sustainability for Bovis Lend Lease in the region.

Rochelle has been engaged with the green building movement for more than a decade, and during that time, she has acquired accreditations in LEED, BREEAM, Green Star, Green Star NZ, Passive House, and Homestar. She also has a Master's Degree in property (explicitly focusing her research on any additional cost premium for Green Star certified buildings in Auckland, New Zealand) as well as a PhD in property, where she studied whether the Homestar rating tool delivered homes with improved indoor environment quality.

Rochelle has worked with the New Zealand Green Building Council (NZGBC), helping it design and implement various Green Building rating tools, including Green Star, NABERS, and BASE. Moreover, Rochelle has also helped the NZGBC design and implement the Homestar residential Green Building rating tool.

Scan the QR code to view Rochelle Ade's Linkedin Profile

Figure 41 Photo Contrasting Established HVAC Solutions with a Growing "Green Wall". Technology vs Nature.

# The Green Building Movement in New Zealand, Australia, and South Africa

In these regions of the world, the Green Star certification program is frequently used. Primarily, Green Star, which is loosely derived from BREEAM and LEED with country-based subprograms, including Green Star New Zealand, Green Star Africa, and Green Star Australia, which have been around for more than a decade.

In Australia, Green Star is utilised for commercial buildings and multi-unit residential setups. However, in New Zealand, it is only used for commercial buildings. For residential buildings, the New Zealand building council has developed its own tool (Homestar), and it states that the Green Star program cannot be used for the residential buildings in New Zealand.

Nevertheless, Rochelle disagrees with this scenario; she points out that Green Star can also be utilised for residential buildings. She contends that, although it cannot be used with single-family, standalone-detached houses, it can still be applied to the apartment buildings. The Passive House standard also has a presence in New Zealand and Australia, and it is gaining greater traction every year.

# Carbon Footprint and Reference Points

Green Building rating tools award points to 'less bad' buildings than brown buildings, and they do not seek to consider what constitutes good and award points accordingly.

For example, the Building Research Association in New Zealand has calculated a carbon budget for new commercial and residential buildings. However, instead of awarding points and rating on good behaviour (i.e. under the carbon budget), they use a 'bad building' benchmark and award points based on how 'less bad' you are than the bad building. We should not be seeking just to be 'less bad'. We should be striving for good, if not excellent.

Figure 42 CO2 Footprint

# Carbon-Based Metrics Distract From the Importance of Water Conservation

Everyone is concerned about carbon, and Rochelle points out that the focus on carbon causes us to ignore various other environmental impacts to our detriment. For instance, in New Zealand, another major issue is the degradation of waterways, but while our primary focus remains on carbon, we are not improving the health of the waterways.

While it is essential to understand what we are doing to our atmosphere, it is also imperative to realise what we are doing to other planetary systems; for example, our oceans and coral reefs. Therefore, we should revisit our primary focus on carbon only and take a more holistic viewpoint, incorporating a broader range of measures that align with the planetary boundaries.

# Green Building Certification Programs

Rochelle recently undertook a grading of Green Building rating tools worldwide using the standard grading scale from the University of Auckland for students' grades. She concluded that a 'world-leading' 6 Green Star rating in New Zealand and Australia was the equivalent of a "B" compared to the "A" for the USA and the UK. This is a problem which questions the relevance of green building rating programs. She stated,

*"I don't like certification systems. I think they should all be abolished quite frankly; I think they provide a false sense of reality."*

Rochelle doesn't believe that we need performance green building rating tools; instead, everything should conform to a basic standard.

Primarily, the buildings should be low energy and comfortable inside. If any building does not meet basic requirements, then it should not be given a certification.

Most green building councils provide a roadmap to transform the building design and construction industry slowly on a journey, but what about the here and now? Many buildings are inefficient and unhealthy right now. With these buildings, what collateral damages are we willing to absorb as a society?

There are good intentions, but the fault lies in their implementation. The industry primarily sets up green building rating tools for the industry. Everyone gets a pat on the back; e.g., the developer receives a plaque to put on their wall, the architect gets a certificate for their design, and the builder gets to claim a green build. Still, there is no check and balance to evaluate the actual performance of these tools, especially to see if the occupants are happy, healthy and content.

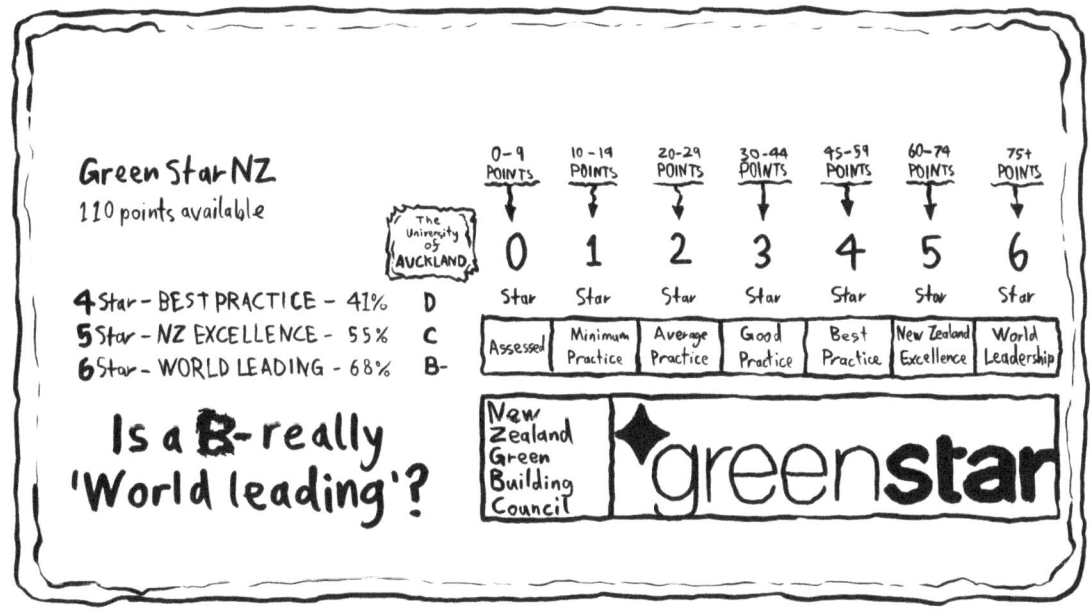

Figure 43 Results from Grading Green Building Rating Tools. Source: Rochelle Ade 2021

# Occupants Matter

To ascertain if something is energy efficient i.e., water-efficient, carbon-efficient, or whatever metric is used, we must analyse the building and its occupants. When a building is constructed, people move in and work in it. Consequently, its performance is a symbiosis between the building and its occupants. However, every green building certification still dissociates these two factors.

The unsatisfied occupants seldom have a voice, and there is no system to give their feedback to the building designers, constructors, developers, and local authorities. If any post-occupancy assessments are performed at all, they tend to be a tick box checklist. In essence, real-time performance over time is what matters, and green building certifications do not address this issue.

# The Way Forward

Rochelle started her PhD as a believer in green building certification programs. However, after performing detailed research, to prove that a green building rating tool (Homestar) delivered houses with superior indoor environment quality she found 'brown' houses, and accidentally proved that they didn't. This changed her worldview.

Consequently, Rochelle now believes that green building rating tools are ineffective. Industry writes them for its own ends with no regard for performance verification. She contemplates that we should relinquish these tools and programs and replace them with something that works.

Due to her views, Rochelle has experienced some pushback from the industry. Notably, the most concerned people are the ones with a vested interest who strongly support the status quo.

# Passive House Bridges the Gap Between 'How' and 'Why.'

Typically we teach people how to do things and not necessarily why; closing this gap between 'how' and 'why' can substantially improve building construction design.

Unlike green building rating tools, Passive House is very clear on its why i.e., to enable the design and construction of warm, dry, comfortable, energy-efficient buildings. It does not dabble in transport or bike racks, and Rochelle's opinion is that it is rigorous and actually delivers warm, dry, comfortable, low-energy buildings.

Scan the QR code to listen to Rochelle Ade's Edifice Complex Episode

# Advice to New Graduates

It is plain and simple, take a Passive House course. You will learn more in two weeks than in your entire college and university studies. As a result, following this course, you will be able to design better buildings.

In addition, think about your passions, your why, and get some work experience without studying right away. Get a job, learn and discover the parts you love, do not start your career without a test drive.

Scan the QR code to visit Rochelle Ade's Twitter

# Reader Notes: Green Building Certification is not Working

Readers are encouraged to note their favourite insights, triggered ideas, and next steps.

# Bill Browning

## Biophgilic Design

I really enjoyed interviewing Bill because he helped me see things in plain sight, that I had missed as I drifted along, not seeing what was in front of me. Bill's approach to using Biophilic design principles and adding real value to the quality of the user experience plus the financial bottom line is refreshing.

The demonstrable benefits of Bill's work leave me amazed that the principles of Biophilia and Biomimicry are not incorporated into the mainstream building and interior design work. I guess this is a testament to the established "Construction Industrial Complex" ability to resist change.

If you are a property developer who retains the built asset and derives income from the building user experience, incorporating Biophilic design principles is one way of gaining a competitive advantage. In my opinion, we need to get the word out about Bill and his work.

*"We still need to do some work... but it's one of the reasons why for us, biomimicry is intriguing because nature's got 3.8 billion years of experimentation out there that we need to learn from."*

Bill Browning is one of the foremost thinkers and strategists of the green building movement and the real estate industry. Bill is a strong advocate for sustainable design solutions at all levels of business, government, and civil society. His expertise has been sought out by organizations as diverse as Fortune-500 companies, leading universities, non-profit organizations, the U.S. military and foreign governments.

Bill helped build Buckminster Fuller's last experimental structure early in his career. Then, in 1991, he founded Green Development Services at the Rocky Mountain Institute (RMI), an entrepreneurial, non-profit "think and do tank." His consulting projects at RMI include the design of new towns, resorts, building renovations, and high-profile demonstration projects. These projects comprise Wal-Mart's Eco-mart, the Greening of the White House,
the National Aquarium, Disney Hong Kong, the Pentagon, Lucas Film, Grand Canyon National Park, and the Sydney 2000 Olympic Village. In 1999, Green Development Services was awarded the President's Council for Sustainable Development/Renew America Prize.

Bill received a Bachelor's degree in Environmental Design from the University of Colorado, specializing in energy-conscious architectures and resource management. Moreover, he also holds a Master of Science in Real Estate Development from the Massachusetts Institute of Technology (MIT), where he was awarded the MIT Center for Real Estate's 1991 Public-Sector Fellowship. In addition, he also attained the Charles H. Spaulding Award in 1995. In 1998, Bill was named one of five people "Making a Difference" by Buildings magazine. In 2001, he was selected as an honorary member of the American Institute of Architects. Subsequently, in 2004, he was honoured with the U.S. Green Building Council's Leadership Award. He is a founding member of the U.S. Green Building Council's board of directors, and in 2006, Bill founded Terrapin with longtime partners Bob Fox, Rick Cook, and Chris Garvin.

Scan the QR code to view Bill Browning's Linkedin Profile

# What Would Buckminster Fuller Have Thought of Property Development and Architecture Today?

Bill has remained a member of the board of the Buckminster Fuller Institute. As a design figure, Buckminster Fuller pushed technology to solve humankind's growth issues, and Bill worked closely with him. That is why Fuller's thinking and work heavily influence Bill. According to Bill, some of the ideas Buckminster Fuller worked on have come to fruition today, and others have not. However, Buckminster Fuller would have been astonished by the development and growth of computing and electronics within the built environment.

Buckminster Fuller pushed the concept of achieving more with fewer resources, which is an underlying objective of sustainability. However, there is still much work to do that biomimicry can achieve. Biomimicry is intriguing because nature has 3.8 billion years of experimentation across the planet, which can be utilized to learn new ideas.

# What is Biomimicry?

By definition, Biomimicry is an innovation inspired by nature. Essentially, it is a process of learning from nature to solve various problems. In this regard, Terrapin ran a program for five years for the New York State Energy Research and Development Authority (NYSERDA), which worked with various companies to investigate problems that conventional engineering could not solve.

Figure 44 R. Buckminster Fuller

# Basis of Biophilia

Biophilia is a combination of 'bios' and 'phileo' or 'life' and 'love'. This term can be attributed to the social psychologist Erich Fromm and Harvard biologist E.O. Wilson popularising it. Biophilia implies how the experiences of nature affect us physiologically and psychologically. In addition, more work is being shifted into the neurological aspects. After evaluating hundreds of research articles and working with various researchers, we have found that nature experiences can be categorized into the following three major brackets:

**Nature and Space:** these are the direct experiences of nature in the built environment. They involve a view of nature, plants, animals, water, water in space, and light and thermal variability.

**Natural analogues:** these are the indirect experiences of nature. They involve using natural materials, biomorphic forms, and fractal patterns, which elicit an interesting brain physiological response.

**Nature of the space;** these are three-dimensional experiences that elicit distinct responses, of which some are relatively well known. Two of these responses are described as follows:

- **The prospect:** Significant for wayfinding and the sense of security; the opposite or coupling of this is the 'refuge'. A classic example of a prospect and a refuge together is a beautiful craftsman bungalow with a big overhanging porch on the front. When you sit on the porch, you are about 18 inches elevated from the ground plane, and here you have a more prominent cone of vision; you can see up and down the street. Hence, you have excellent prospects. Moreover, the wall also protects your back, and you have the overhanging porch over you to have a great refuge space. Hence, you have both prospect and refuge together.

- **The mystery:** In this case, you have partially revealed information, and you feel compelled to explore further. It is just like taking out your kids and dogs for a walk in the park, then you reach a curving path, and they bolt ahead to see what is around the corner. With a strong mystery condition, you experience a sense of compulsion to experience it. Many bakeries use this idea: you catch the scent of cookies down the street, and you want to explore its origins. This is a risk peril and is a pattern that should not be used too much; it elicits a strong dopamine response in the brain with possible danger.

## Why are the Philosophies and Principles of Biophilia and Biomimicry Not Mainstream?

It seems that these philosophies and principles have simply been forgotten. However, we can observe that some interior design and architectural programs incorporate Biophilia principles. The design community is starting to shift toward Biophilia, and the new headquarters of The American Society of Interior Designers is a solid and explicit exercise in Biophilic design.

The society is now studying how its staff responds when entering the building to gather behavioural feedback on the impact of Biophilic design outcomes.

## Biophilic Design and the Guest Experience in Hotels

Research was undertaken to determine if Biophilic designs change the guest experience in hotels and influence the economic performance of hotels. The answer was yes! The observational study consisted of six hotels in midtown Manhattan, where three of them had plain standard lobbies, and the other three had explicit Biophillic designs. The study evaluated the ratio between the people who were just transitory and people who were spending time in the lobby. It concluded that a 36% increase was observed in the number of people spending time in the lobbies with Biophilic design measures. This implies that there is a 36% more chance that they will buy a drink, coffee, or snack to increase the hotel's revenue.

Scan the QR code to listen to Bill Browning's Edifice Complex Episode

# Examples of Fractals in Building Construction

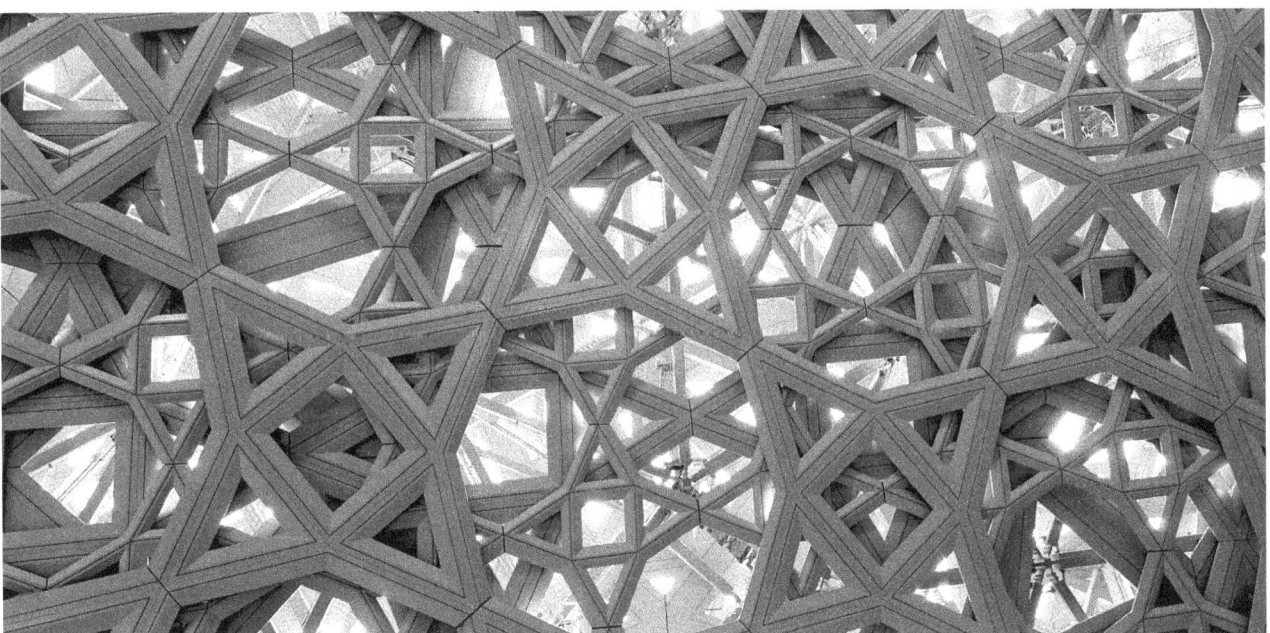

Figure 45 Lourve Abu Dhabi

Figure 46 Lourve Abu Dhabi Roof

## Hotel Lobby Design and Fractals

Fractals can be introduced via perforated metal screens, fabrics, wallpaper, and natural materials like wood grain. There has to be a level of care for fractals, especially when it comes to iterations. For instance, when we take a single square, we make a checkboard with six squares, which is the first iteration. When we perform this again, and for a third time, we can get a Coptic cross or a cross-stitch pattern. Furthermore, it becomes something spiky and scary if we do it in the fourth and fifth iterations. Hence, this signifies the process of creating fractals concisely. Not enough is boring, and too much is scary. Overall, a design created on the third iteration will be preferred. In nature, fractals occur in natural elements like snowflakes and fern leaves.

We are naturally fascinated by fractals, and the neurosciences say fractals are a bit like brain candy. One of the theories is that when we see fractals in the built environment because they also occur in nature, things our brains are already attuned to them. It takes less energy for the brain to process them, making them a pleasurable and low-stress response.

## Attention Restoration Theory and Suboptimal School Design

School design and construction are already as low-end as possible with next to no innovation. It is ideal for exploring Biophilia's human connection to nature in classrooms, considering its low-end design and construction outcomes.

It is worth noting that it is not ideal to have uniform thermal and lighting conditions throughout a classroom. We did not evolve in an environment with consistent thermal and light conditions as a species, so we want some variability in both. In the 1970s, in North America, many schools were built with no windows, with the idea that windows are *'distracting'*. However, it is now well researched that if we look away at nature during a class and change the way the brain is processing, the prefrontal cortex quiets down briefly.

Consequently, when the prefrontal cortex is reactivated, we have a significantly better cognitive capacity; this is called 'the attention restoration theory.' Another relevant question could be: 'How long do we need to look out at nature for this to happen?' This was researched, and the University of Melbourne published an article answering this; all it takes is merely **40 seconds!**

# Eleven Degrees: A Biophilic Design Measure and its Impact

Attention restoration entails the capability to improve our cognitive function in a classroom or office. An example of this was a call centre (office) for a utility company. A scenario is that people are calling in, and not everyone is happy, and they have a problem with their bill, or maybe their power is off. This call centre is a LEED gold building with excellent daylighting, good air quality, raised floor, and good thermal performance. There are trees and a field outside, but because the desks are perpendicular to the windows, people answer calls, look at their computer screens, and miss the view outside—the outside view not even in the peripheral vision.

The intervention was to move these workstations 11 degrees off perpendicular to put the outside view within the peripheral vision. When the leaves outside start to move or a bird or butterfly flies past, it captures attention. As a result, the person looks up, and if he can observe these natural phenomena for **40 seconds,** their brains are reset, which can quiet down their prefrontal cortex. Subsequently, the person is power focused again with a better cognitive capacity.

This call centre intervention resulted in a 6% increase in call handling capacity, which was a return of about $3,000 per desk. This positive return on investment is why many tech companies invest in Biophilia in their workspaces to reduce stress and improve cognitive function among their employees. This is why Facebook, Google, and other tech giants have designed their office buildings with strong Biophilic design measures.

# Concrete Jungles

New York has a regular grid. Therefore, implementing Biophilic solutions within engineering and architecture comes down to the choice of materials. For instance, how light and air are being brought into the buildings, interior design, and people's movements into and throughout the building.

There is currently a movement to occupy roof space rather than giving it over to mechanical systems. The focus is on creating habitat on rooftops with commercial agriculture. Companies like Brooklyn Grange are farming their rooftops, and the success is extraordinary. An above-ground train line, The Highline, has been repurposed into a public walkway with lots of wildlife and artefacts, and it is attracting about 3.8 million visitors annually.

Worthy of note, Tim Beatley from the University of Virginia School of Architecture founded a group called the Biophilic Cities Network. More than 20 cities around the world are now part of this network. Each of them is exploring various mechanisms to connect their citizens with nature and the experiences of nature in the fabric of the cities. Singapore, for example, has a system for recording and measuring biodiversity and the green area ratio mechanism, and they shared their experiences with other cities through the network.

## Advice to New Graduates

Pick a topic you love, chase it to the ground and find mentors; this applies to all the engineers and designers who want to design something different and not be beaten down by the system.

As for advice to the architectural students, it is suggested that they take some time while in school to play around with Biomimicry and Biophilia. They should look at these new topics and not be afraid to ask any questions."

Figure 47 The Highline, New York City

Scan the QR code to visit
terrapinbrightgreen.com

# Reader Notes: Biophillic Design

Readers are encouraged to note their favourite insights, triggered ideas, and next steps.

# Bill Gnerre

Why Does It Take So Long to Set Up BMS?

Bill asks and answers the question that I have heard in every country I have worked in, why does it take so long to set up the building management (automation) system? Suppose you see a stressed-out project manager on a new construction project worried about being able to hand over a fully operational building. In my experience, the cause of that worry usually is the controls systems testing and commissioning.

Bill is a practitioner in a world that is genuinely lacking experienced practitioners, and Bill also has the data to back up his thinking and conclusions. Bill and I agree about his observations on the lack of standard building type operational strategies and the utter lack of standardisation or rigour applied to the systems automation process.

Bill is irritated that today's building systems automation delivery process is 20 years old and needs updating. So am I.

*"The purpose of control systems is to deliver comfort, health and safety in every space."*

Bill Gnerre has a Bachelor of Science in Mechanical Engineering from Northeastern University, and he is the CEO and Co-founder of Interval Data Systems Inc. (IDS). In 2017, IDS was selected by the United States DOE's Smart Energy Analytics Campaign as the Outstanding Monitoring Based Commissioning Provider. At IDS, Bill and his team have one primary goal, *"to operate buildings in conformance with comfort and health & safety standards, at the lowest operating cost."* To achieve this goal, Bill and his team at IDS have developed one of the first enterprise energy management platforms called EnergyWitness ™.

IDS has analysed operating data from hundreds of buildings and has over 800 building-years of BAS data and over 10,000+ building years of building energy data. In 2022, IDS received ASHRAE's 1st place technology award as a key member of a data-driven 80 building retro-commissioning project.

Scan the QR code to view Bill Gnerre's Linkedin Profile

## Purpose

According to Bill, the purpose of a control system is to deliver comfort plus health and safety in every space. This means that the control system, controls and optimises temperature, humidity, ventilation, lighting, safety systems, access control, and energy management.

In Bill's opinion, there is too much focus on energy saving at the expense of comfort and safety. Moreover, this focus is driven by the high visibility of rising energy costs. However, energy is only one of four cost inputs, and the other inputs are maintenance labour, equipment service life, and systems recommissioning / loop tuning/optimisation.

## "Sit-Rep"

Understanding building control systems can be simplified into two different parts. One is, why does something expected and unexpected happen? The other is to ask the following question: what can be done to fix an issue? To answer these questions requires a controls firm to provide software, hardware, and systems engineering expertise.

Understanding systems performance and why issues occur comes down to the controls and automation software used to run the building systems. This software has to be written and implemented consistent with the design intent. A common weak point in practice is that the controls sequences of operation prepared by design engineers are generic and frequently not precise.

The system designer and engineering firm are responsible for producing, reviewing and approving the systems controls sequence of operation. The approved systems controls sequence of operation based on the owner's project requirements is the basis for controls system design, installation, and commissioning. Moreover, it is also essential that facilities managers update the systems sequence of operation when any system modifications are implemented. Systems that work and generate building operational efficiencies begin and end with the system controls sequence of operation.

Another weakness is in the way building controls software is written by the local technician who does not follow common software delivery procedures. It is common for technicians on-site and under time pressure to copy code from a previous project and adapt it for the new project, then test it in the field. This is sloppy and lacks testing and debugging.

The reality in the field is that software engineering is being performed "on the fly" by controls firms, which are supposed to be hardware firms. This is one reason why the process of setting up and handing over building control systems is frequently chaotic and behind schedule.

## Sequence of Operation

The controls sequence of operation, also known as a control's logic statement, is the system operational sequence and controls logic expressed as a narrative plus a controls points list and a schematic diagram. Sequence of operation is systems-based and covers all possible operation modes from start-up to shut down. They are among the most important design and commissioning documents for any project. Without a proper well written, unambiguous sequence of operations, the controls programmers are left to make decisions independently.

A significant problem commonly encountered with control systems is the sequence of operations. They are either over-complicated or generic to the point of uselessness, and sometimes they are simply, absent.

# Operational Data

As technology advances and data storage costs fall, building systems data collection and analysis will become an increasingly larger part of the controls firms' business. In the future, we may also witness disagreements over who will own, collect, analyse, sell, and benefit from the vast volumes of data collected by the controls system.

Bill has experienced controls manufacturers demand that their devices are not tampered with and that the manufacturer owns the data they collect. With the advent of 5G and the Internet of Things (IoT), controls devices on the market, e.g., Belimo, transmit data as soon as they are activated. I believe that we are still in the early days of a new IoT enabled building control paradigm. In the future, IoTc enabled controls systems will commission and tune themselves once installed and powered up.

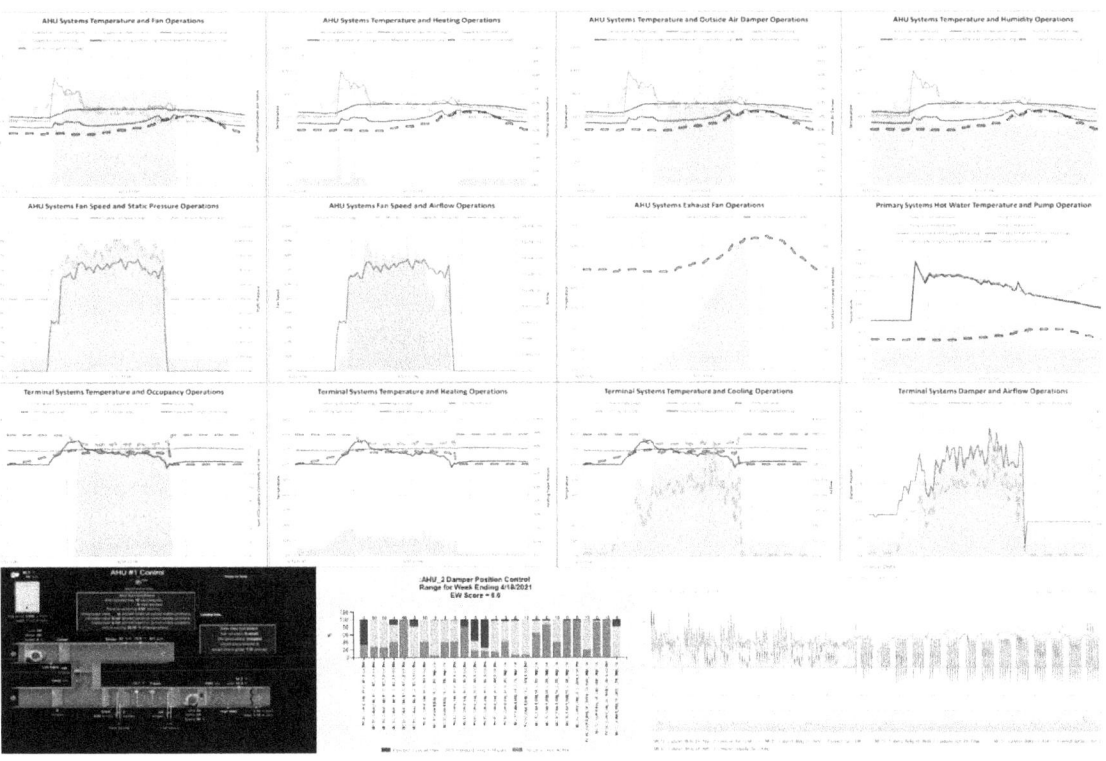

Figure 48 Example View of Building Operational Data Dashboard. Source: IDS 2022

# Integrated Systems Testing

The integration of systems and subsystems is another area that cannot be ignored. In the past, a "silo" approach had been sufficient, and today, the complexity of building systems and their high technology subsystems require interoperability. Bill believes that detailed systems integration testing will become a necessary process for commissioning a building and obtaining the 'Authority Having Jurisdiction' approval for occupancy.

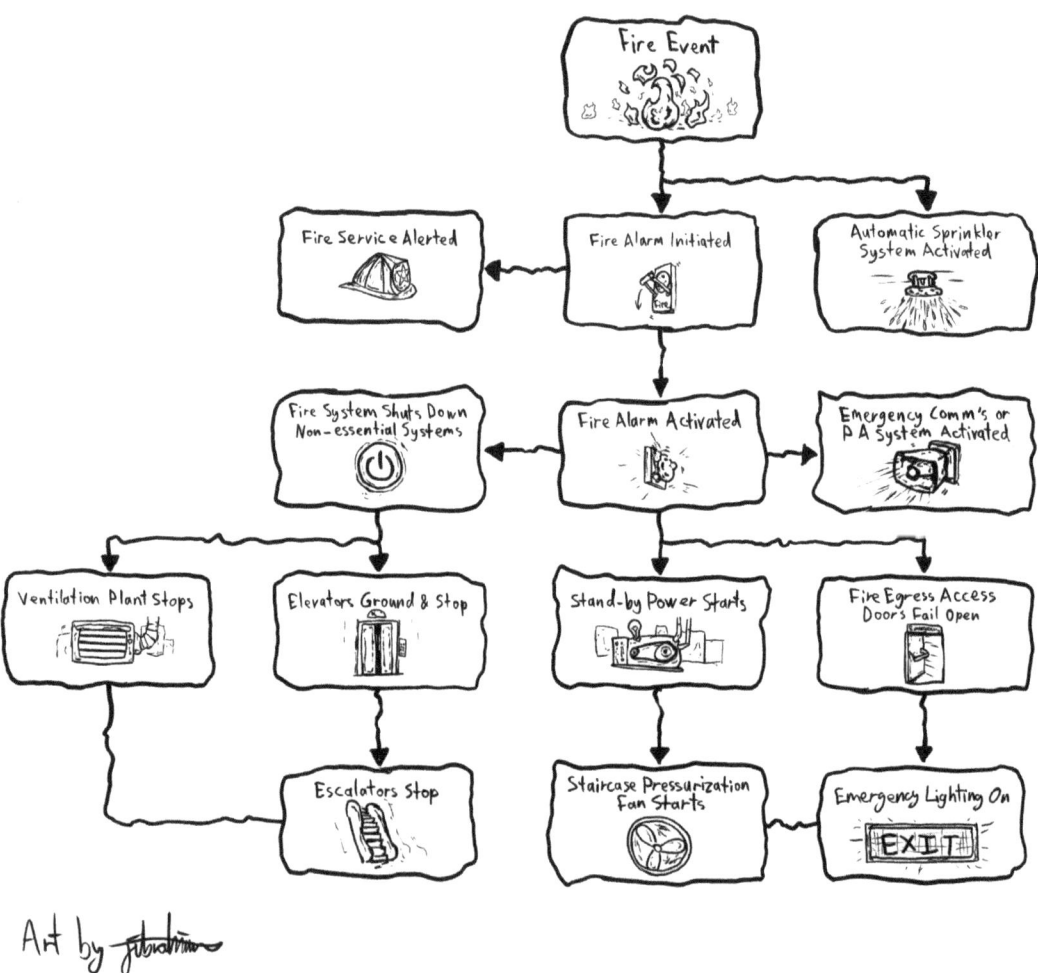

Figure 49 Example Cause & Effects Diagram used for IST Testing. Source: Building Whisperer 2022

# Facilities Management

Facility and building managers will increasingly require controls systems automation expertise and detailed applied systems knowledge. Moreover, training and post-occupancy support will become increasingly significant. Typically, a building management team will outsource controls systems maintenance to a specialist firm in a sub-contracting arrangement. However, skilled and experienced controls technicians are available during occupancy hours to support mission-critical functions. I believe that the large facility management companies will buy small specialist controls firms to bring this expertise in-house.

Training and post-occupancy support will also become increasingly significant. Typically a building management team will outsource controls systems maintenance to a specialist firm in a sub-contracting arrangement. However, to support mission-critical functions, there need to be skilled and experienced controls technicians available during occupancy hours. Large facility management companies will likely buy small specialist controls firms to bring this expertise in-house.

Scan the QR code to listen to Bill Gnerre's Edifice Complex Episode

Scan the QR code to visit intdatsys.com

# Reader Notes: Why Does it take so long to set up BMS?

Readers are encouraged to note their favourite insights, triggered ideas, and next steps.

# Murray Guy

## Lean Construction

In a copy and paste world, Murray Guy is original, passionate and consistent. He puts into practice what he preaches and is on a mission to improve how buildings are designed and constructed by physical example and teaching.

This makes Murray a rare person; he is a teacher that deeply understands his subject based on a lifetime's work plus research. Importantly, Murray is a systems and macro thinker; he does not operate in the silo's the property design and construction industry try to keep him in as an engineer.

The construction industry is historically slow to change, and in my darker moments, this industry attitude gets me down. However, when I interact with people like Murray, I become optimistic because his positive attitude and openness to change through improvement and excellence as a first principle also make me want to change things. The right leader and teacher can really change your life.

*"character, and I call them explorers wanting to make the change and a difference in the industry."*

Murray Guy can best be described as an entrepreneur with a mindset for sustainability. Owning and operating three different companies actively involved in creating a more efficient Green Building industry has afforded him the practical opportunity to demonstrate how lean practices can be used in constructing NetZero buildings.

Murray serves on the Lean Construction Institute Canada board, where he works to build lean project delivery capabilities, develop the certification program, and establish communities of practice. He is the CEO of Integrated Designs, where, together with his team, he works to deliver high-performance buildings. At EcoSmart Developments, he designs and builds NetZero ready homes, and at Shift2Lean, he offers training and coaching on Lean Project Delivery Methods and Tools.

# Integrated Design Process and LEED

The Integrated Design Process (IDP) is a much-lauded but seldom used concept. Actual integrated design practice involves all design team members working in parallel and coordinating the design work as they progress. Moreover, IDP places equal value on each discipline and specialists involved in the design process. Importantly, IDP presupposes that the architect, engineer, constructor, and commissioning specialist have equal value within the project team.

The LEED green building certification program places high value on IDP as a necessary factor in the design process. Whilst working on a LEED gold project, Murray realized the importance of seeking value and low energy solutions in everything from the design process all the way through to the end of construction. This means enrolling the contractors during the tender phase in high value, low energy solution workshops. The typical silo approach to construction needs to change.

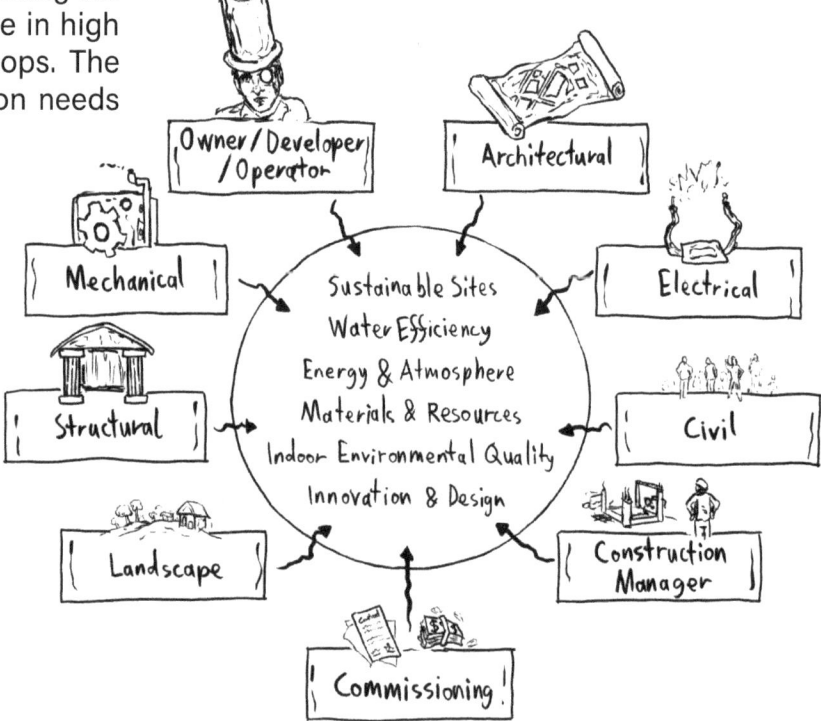

Figure 50 LEED Integrated Design Process. Source: Building Whisperer 2022

# Six Sigma vs Lean

Six Sigma was developed with a particular goal, i.e., reduce variation and defect rates in the production processes through statistical analysis. To do this, Six Sigma utilizes one of two 5-step approaches, either the DMAIC or the DMADV method. Both methods have their specific utilization.

DMAIC stands for Define, Measure, Analyze, Improve, and Control. This process involves identifying the problem you're trying to solve, taking stock of your current processes, identifying and implementing a solution, and maintaining that solution in the future. This is ideally suited for supply chain performance issues.

DMADV stands for Define, Measure, Analyze, Design, and Verify. The first phases of the project are the same, but the design phase allows for creating an entirely new tool to solve the problem. The "Verify" phase then focuses on ensuring that the new solution does, in fact, continue to solve the problem.

Scan the QR code to listen to Murray Guy's Edifice Complex Episode

Six Sigma is all about monitoring the supply chain for defects, identifying issues, and solving them as effectively as possible.

Lean construction methods are entirely focused on eliminating waste, providing maximum value to the customers with the lowest possible investment. Moreover, it involves every tier of an organization or project, helping guide new processes and drive the resource allocation process.

The primary difference between Six Sigma and Lean approaches is that Lean is less focused on manufacturing (i.e., construction) but often shapes every facet of a business.

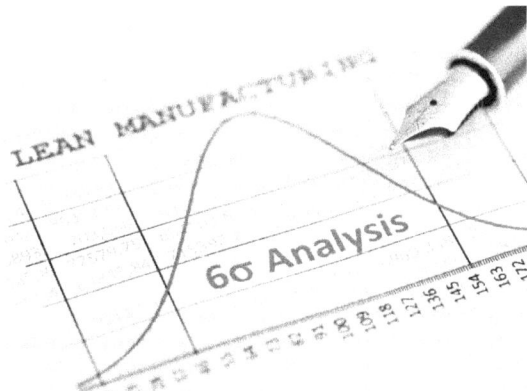

Figure 51 Six Sigma vs Lean

Scan the QR code to view Murray Guy's Linkedin Profile

Wisdom of the Property Crowd | 121

# Construction Management

Construction Management (CM) is a professional service that offers various effective services to the project owner, such as managing the project's schedule, cost, quality, safety, scope, and operation. Construction Management is compatible with all project delivery methods. No matter the setting, a Construction Manager's responsibility is to the owner and a successful project.

A property development project is made up of three parties, excluding the Construction Mangager:

i. The owner.
ii. The architect and engineers who design the project.
iii. The contractor, who oversees day-to-day operations and manages subcontractors.

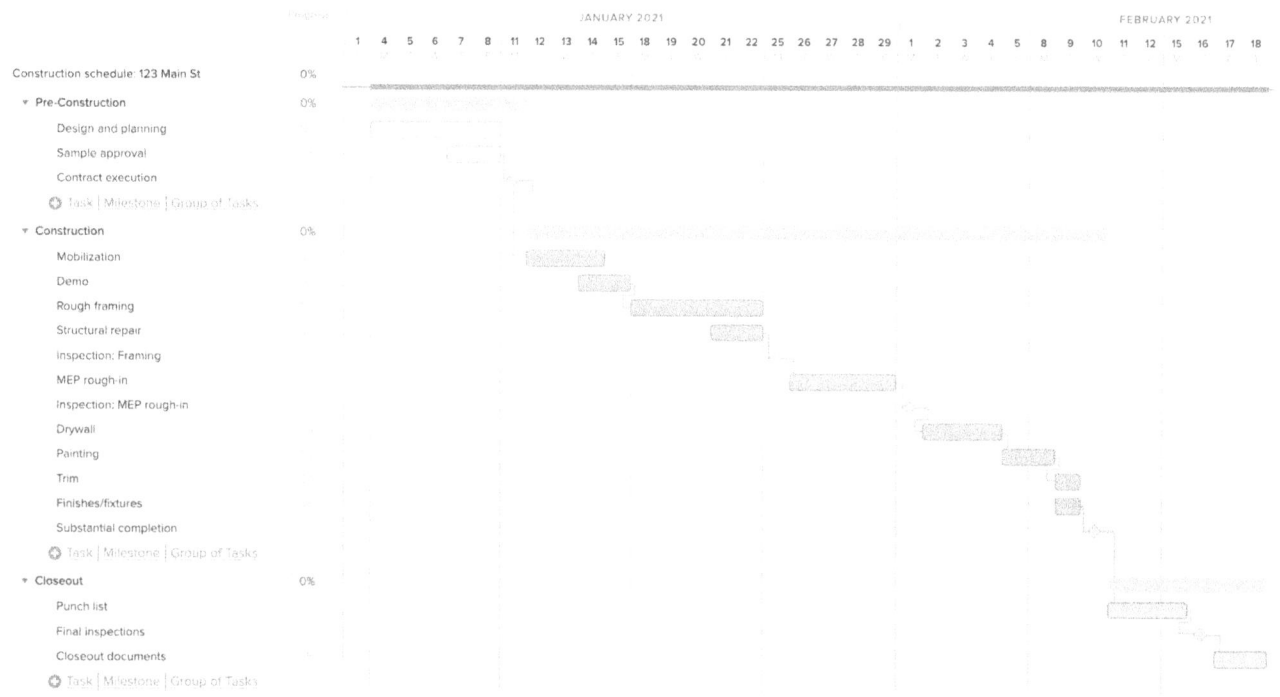

Figure 52 Construction Gantt Chart

The Construction Manager represents the owner's interest and provides oversight over the entire project directly for the owner. His/her mandate is to work with all parties to deliver the project on time, at or under budget, and to the owner's expected standard of quality, scope, and function.

Construction Management is a really flexible approach that allows owners to make adjustments while designing and during early construction. In Murray's experience, Lean projects tend to be delivered by construction managers.

The combination of an Integrated Design Process + Construction Management + a Lean Construction Approach = the best value, lowest price project outcome.

## Why Use a Lean Construction Approach?

In Murray's experience, a Lean Construction approach can yield 18% potential savings in the design process and a figure of about 40% when proper waste elimination is incorporated with Lean in the construction process. But Murray firmly believes these savings can be higher as Lean Construction methods are refined. Even taking a pessimistic view, savings of 20% can be expected from a Lean design and construction process.

Despite these compelling numbers, there is resistance to the property industry's adoption of Lean Construction methods. Murray believes this is due to perceived threats to people's jobs, established supply chains, trades unions, and the property industry's epic history of resisting change.

Unless there are a significant supply chain shock or materials price increases due to shortages, the move to Lean Construction methods will be an evolution rather than a revolution.

## Affordable NetZero Passive Residential Houses

Murray is walking his talk, and he has a business building eco-smart houses that meet a NetZero standard. Crucially, he is offering these options with the target of making the future-ready houses at no cost premium to current market standards. With Lean Construction methods, there can be ample savings on cost, thereby ensuring that residential houses designed and built using lean concepts remain entirely affordable while retaining high quality plus high-performance outcomes.

Scan the QR code to visit Murray Guy's Website
EcoSmart

# Reader Notes: Lean Construction

Readers are encouraged to note their favourite insights, triggered ideas, and next steps.

# Professor Roland Clift

## Ethics

Professor Roland Clift challenged me when we interviewed him. He asked me if I would have worked on the construction of the Nazi death camps. My first and the correct answer is no! Easy for me to say no, from the comfort of my home, not under the threat of violence. However, many ordinary, everyday people, engineers, architects and tradesmen did. Is it OK if you only reviewed a drawing for an hour?

These are questions about personal and societal ethics. Ethics matter because they speak to a duty of care, personal integrity and national culture. Ethics are a part of our core values, and when addressed in aggregate, ethics can dictate a nation's destiny.

Prof. Clift really made me think deeply about ethics, particularly when applied at the macro level when addressing sustainability and climate change. It turns out there are ethical aspects to our everyday choices and consumption.

I am thrilled we had the opportunity to interview Prof. Clift; he has vast cross-disciplinary knowledge and is a thought-provoking educator.

*"It isn't a defence to say, 'I was only following instructions or taking orders.' So, if you're asked to do something as an engineer, and you know that this is going to have negative impacts on people, maybe even a lot of people, then you've got to say no. Ethics is basically about your responsibilities to other people."*

Professor Roland Clift BA, MA (Cantab.); PhD (McGill); CBE, FREng, FIChemE, FRSA, Hon FCIWEM is a Chemical Engineer who helped develop the discipline of Industrial Ecology. In addition, he has acquired the following positions:

- Emeritus Professor of Environmental Technology and Founding Director of the Centre for Environmental Strategy, University of Surrey
- Past Executive Director and President of the International Society for Industrial Ecology (ISIE)
- Visiting Professor in Environmental System Analysis, Chalmers University, Gothenburg, Sweden
- Adjunct Professor in Chemical and Biological Engineering, University of British Columbia, Vancouver
- Visiting Professor in Centre for Industrial Ecology, University of Coimbra, Portugal

Moreover, Professor Roland Clift was one of the first to define and examine *"clean tech"* as a sector within the built environment. In addition to being an academic and educator, he has advised industry and governments using a systems approach to analyse sustainability issues and, importantly, has considered the implied ethical issues.

From 1996 to 2005, Roland served as a member of the Royal Commission on Environmental Pollution (RCEP). He is a former member of the UK Eco-labelling Board, of the Science Advisory Council of the Department, Food and Rural Affairs (Defra), the Royal Society/Royal Academy Working Group on nanotechnology and the Working Group, which drafted and updated the BSI/Defra/Carbon Trust standard on carbon labelling, PAS 2050. In 2004-5, he acted as Expert Adviser to a House of Lords Select Committee inquiry into energy efficiency. His research is concerned with environmental management and industrial ecology approaches, including life cycle assessment and energy systems.

Scan the QR code to view Professor Roland Clift's Wiki Page

# Post-Normal Science

Silvio Funtowicz and Jerome R. Ravetz developed the idea of post-normal science in the 1990s. It involves subjecting ideas and proposed decisions to public debate amongst a particular community when available data and background information are either insufficient or highly uncertain. In practice, it can form part of planning and technical decision-making to ascertain that proposed ideas and solutions represent the wishes of the people who will be most affected by the final decisions.

Prof. Clift believes that this approach can help deal with the reactionary outbursts on climate change, environmental degradation, and urban air quality.

# Economics of Externalities and Cost

Neoclassical economics (market-driven pricing based on supply & demand) makes it hard to debate economic effects on society. The fundamental problem with neoclassical economics is reducing most factors involved to a single dimension called value. The long-term effects and externalities of aggregate human activity leading to climate change and environmental degradation cannot be easily tackled in a neoclassical economy due to its focus on current value rather than costs tomorrow.

With macro issues, such as climate change and environmental degradation, there is also a "tragedy of the commons" trap, i.e., it is everyone's and no one's problem at the same time. This scenario leads to an overwhelming and individual disassociation due to the sheer magnitude of such issues.

By utilising ecological economics, Prof. Clift believes that there is a better approach to discussing macro issues, such as climate change and environmental degradation. This is an interdisciplinary field of academic research addressing the interdependence and co-evolution of human economies and natural ecosystems, both intertemporally and spatially. By treating the economy as a subsystem of Earth's larger ecosystem and emphasising the preservation of natural capital, ecological economics can be distinguished from environmental economics, which is the mainstream economic analysis of the environment.

## Be Proactive

Concerning climate change, clean technology, and sustainability, Prof. Clift believes that the correct focus is to "get it right" rather than "clean it up." During his time at the UK Initiative for Clean Technology, Prof. Clift noted that instead of focusing on cleaning up, they focused on getting it right from the onset. When the general cost of being reactive is compared to being outright proactive, the latter pales compared to the former.

However, it seems that there will always be attempts to defend the status quo, especially by the biggest beneficiaries of the soon to be defunct system. Moreover, path dependency has a long tradition of lobbying and government capture. Therefore, politics will always be a factor when enacting change at societal levels.

# Ethics in Engineering

Ethics matters in engineering, as they deal with bribery, corruption, health & safety, and public and professional duties of care.

It is, in fact, no longer a sufficient defence to say, *"I was only following instructions or taking orders."* Ethics must be incorporated into the everyday life of an engineer. If an employer demands that you, as an engineer, carry out a task, it is your responsibility to ask yourself if it is ethically correct to do this task. Especially when it comes to doing tasks that would affect other people's well-being and safety. Hence, ethics is a matter of personal integrity and responsibility.

Scan the QR code to listen to Professor Roland Clift's Edifice Complex Episode

Figure 53 Thinking about Ethics

# Marketing and Environmental Ethics

Aggregate consumer demand is a powerful force that drives economies, companies' profits, resource consumption, and environmental degradation. Public companies have a fiduciary duty to maximise their shareholders' returns. Consumers demand low-cost goods and services. However, incentives can be misaligned. The environment, not the consumer or profit-seeking companies, pays the price of unaccounted externalities.

Marketing environmental issues to consumers seems to be the best way to drive companies to comply with the ethical guidelines. Companies and governments will eventually listen if consumers demand information on goods sold about ecological impacts and recycling. Thus, the power of the consumer purse is one of the world's most powerful and concealed forces for transformation.

For instance, a marketing drive by supermarkets pushes the interest in environmentally friendly products, which in turn motivates the consumers to become more conscious of what is on the labels of their purchased products. Furthermore, the marketing mixes of these supermarkets are inclusive of environmental credentialing, which effectively creates a culture of buying environmentally friendly products. In essence, marketing can drive the culture.

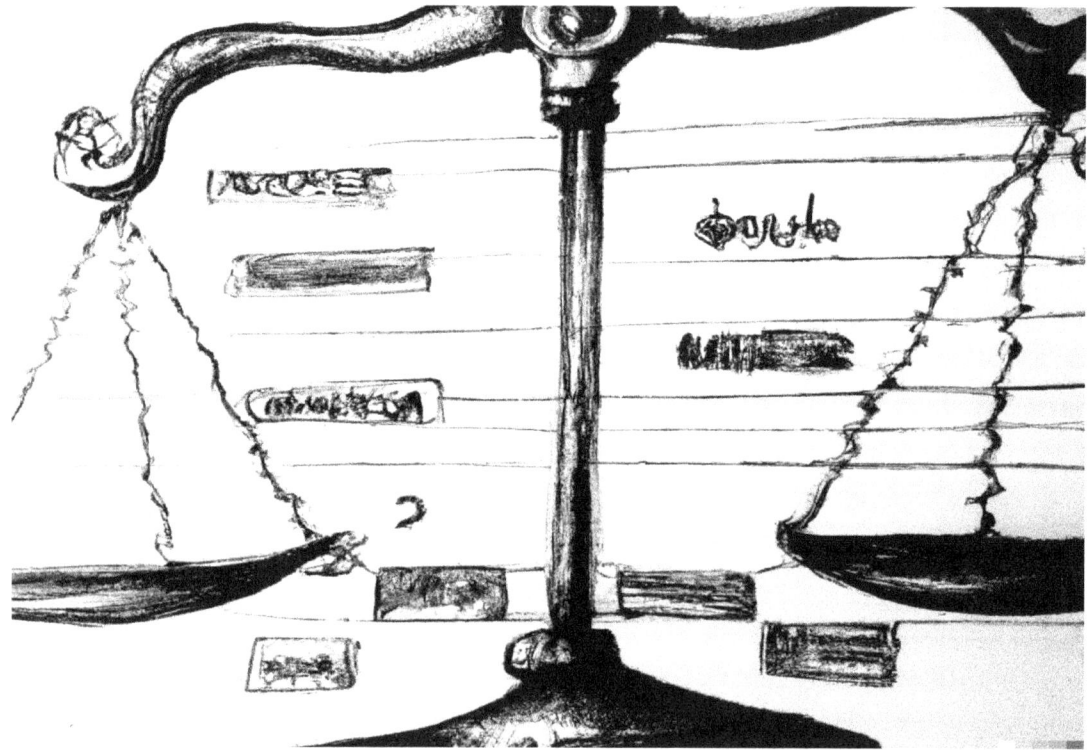

Figure 54 Ethics

# The Venn Diagram of Sustainability

The three-lobe Venn diagram below is a pictorial explanation of sustainability, which is commonly misunderstood.

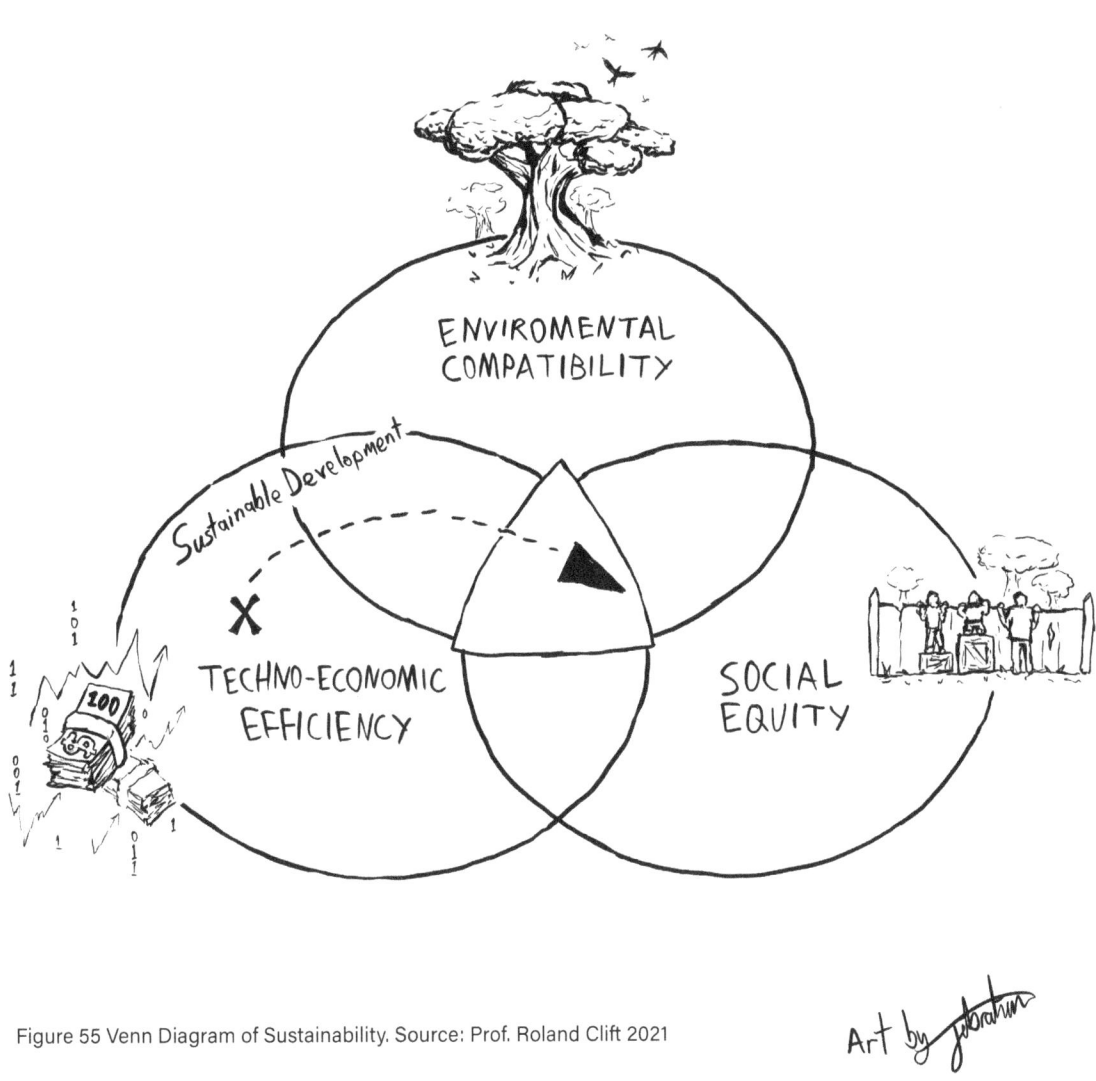

Figure 55 Venn Diagram of Sustainability. Source: Prof. Roland Clift 2021

According to Prof. Clift, the Venn diagram explains decision spheres. Maintaining consistency with one space would effectively lead us to the other space, as all of these three spheres are interdependent. In essence, sustainable development exists at the intersection of the three lobes.

## The Paris Agreement

While the 2015 Paris Agreement (popularly known as the Paris Climate Accords) and their successors play a significant role in tackling the climate change issues, it is a "top-down" intervention with no real compulsion unless enacted into law by the signatory countries.

Prof. Clift believes that more impact would be experienced and progress made if the sustainability movement is a grassroots phenomenon. It is more accessible to develop momentum and generate change through local actions and choices with a local foundation. Moreover, scalability is also more achievable when a movement has an organic drive. It comes from the people rather than through attempted enforcement from the governmental authorities and legislative processes.

## Three Things to Change

In picking three factors that he would like to change to make a difference, Prof. Clift takes education as his first pick. According to him, he would put Industrial Ecology into engineering systems' education and practice. This is because it covers systems thinking, ethics, environmental concerns, equity, and analysis. His second pick would be changing the stranglehold of economics on public policy. He disapproves of the short-termism that arises from the emphasis placed on discount rates. Finally, his third pick is to incentivise the financial markets, so the investment sector takes a longer-term view.

Scan the QR code to visit Professor Roland Clift's Website

# Reader Notes: Ethics

Readers are encouraged to note their favourite insights, triggered ideas, and next steps.

# Jerry Yudelson
*The Godfather of Green*

One of the built environment's issues in attracting talent is the lack of visible role models and "rock stars". In my opinion, Jerry Yudelson, the Godfather of Green, ticks this box. As an early champion and leader in the sustainability and green building space, Jerry defined what it meant to be a "green leader".

Jerry has influenced thousands of people through his work with the U.S. Green Building Council and the training he has delivered. Significantly, Jerry stepped up to be a public face for sustainability and a communicator between industry professionals and the public at large.

It would be easy for Jerry to sit back and take it easy, trading on his past body of work, but he is not doing that, which is why he is a role model for me. He is still curious, passionate and forward-looking in his work.

Jerry Yudelson is a leading American expert on green building and has degrees from Caltech, Harvard, and the University of Oregon. An author of fourteen books, Jerry is a highly rated keynote speaker on sustainability and green building, having spoken at more than 100 green building conferences in 20 countries. In 2011, *Wired* magazine anointed him as "The Godfather of Green."

With over twenty-five years of experience under his belt within the field of renewable energy systems and green building design, Jerry is a keynote speaker at major building and sustainability events. His engaging 2020 memoir, *The Godfather of Green: An Eco-Spiritual Memoir,* shares many stories from the early days of the green building movement.

Scan the QR code to listen to Jerry Yudelson's Edifice Complex Episode

## LEED

Jerry was one of the first ten faculty members of the U.S. Green Building Council (USGBC), where he trained over four thousand people. Jerry was named a LEED Fellow in 2011. The USGBC and LEED building certification program started in 2000 with the best intentions but after twenty years seems to be as much a triumph of marketing as a substantive change agent for the building industry.

Seeing things from the inside, Jerry saw issues with transparency and openness concerning the actual performance of LEED certified buildings. One big problem is that actual building performance was never reviewed or verified after a LEED award, making sustainability claims and building performance rather uncertain. After 2012 Jerry witnessed a stall in the growth of LEED certification. This was due to the failure of LEED certification to break out at scale into a commercial property outside major initial markets and a complete failure of its entry into the residential property market, along with resistance from professionals and industry players and "certification fatigue."

Scan the QR code to view Jerry Yudelson's Linkedin Profile

Over time, the LEED certification program became complex and challenging to navigate. For example, by 2015, there were more than 11,000 interpretations of LEED credits within the USGBC database. Also, the program is expensive and time-consuming, and many major players began to abandon it.

Despite the continuing interest in sustainability and "green buildings", Jerry believes green building certification needs massive changes, a process he outlined in his 2016 book, Reinventing Green Building. This reinvention needs to simplify and automate the certification process as much as possible. There are currently over 600 green building certification systems worldwide; if LEED does not reinvent itself and compete, it may become irrelevant.

Figure 56 LEED Certification's Decline. Source: Jerry Yudelson

# Indoor environmental quality & energy performance

Jerry believes the future of building performance and maybe green building certification should be heavily tied to indoor environmental quality (IEQ), something reinforced by the 2020-2022 pandemic. IEQ refers to the quality of a building's environment in relation to the health and well-being of occupants. IEQ is determined by many factors, including lighting, air quality, off-gassing of contaminants, ventilation, temperature and humidity conditions.

The optimum confluence of design factors is the balance between meeting IEQ whilst minimizing energy consumption and environmental degradation. However, there is uncertainty and a lack of agreement about what to measure for IEQ plus; energy metering and submetering are not common.

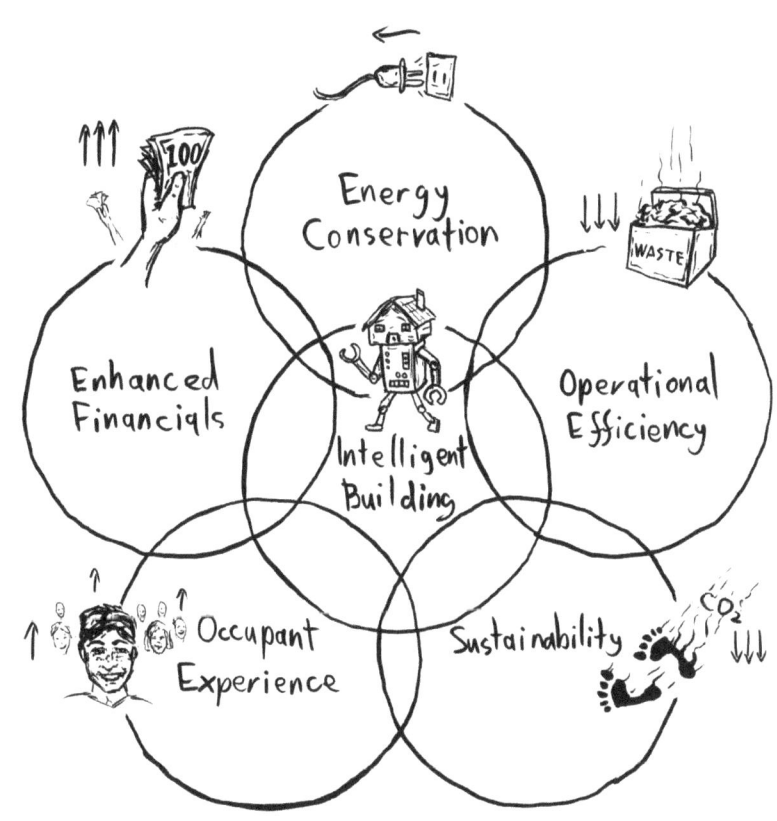

Figure 57 Building Operational Verifications. Source: Jerry Yudelson

Occupants' well-being and satisfaction can be sampled via real-time online polling. Compiling polling results with building systems performance data over time can enable practical trend analysis for facilities management and spur timely corrective action.

Any useful building certification system (LEED, BREEAM, WELL, Green Star etc.) must be referenced within the building management systems, incorporate detailed IEQ, Energy strategies plus verification, timely annual reporting on operational efficiencies, user experience and actual costs.

## Culture & Geography

Cultural and geographic context really matters in building design. A one-size-fits-all, rule of thumb design approach is a recipe for poor performance and occupant satisfaction. However, if left unchecked, cost pressures drive design to a one-size fits all, rule of thumb solution set. For sustainable, green buildings, a minimum requirement is a detailed design that acknowledges cultural and geographic factors.

Taking a considered, detailed design approach is not all bad news. There are always opportunities for free cooling, free shading, wind, and solar energy. Detailed design forces consideration of all the natural geography can offer. Furthermore, some cultures do not like enclosed spaces maintained at 21 degrees celsius with full air conditioning. Factoring in local culture and living habits can reduce energy load considerably and enhance occupant satisfaction.

The management of and interaction with professionals and constructors in various cultures can also influence a project's outcome. The design must take care of local norms, customs, and sensitivities then apply specialist knowledge or innovation in a "gentle" fashion. This implies that the Green Building strategies must have undeniable benefits, and your persuasion skills must be accurate and to the point. In Jerry's opinion, the integrated design process is the best way to achieve an optimum design solution.

# It is Either Net Zero, or It Isn't

There are several definitions of NetZero, and over time the property industry needs to find one definition everyone can agree on. For example, in the U.K., NetZero Carbon is a definition that the property industry seems to be targeting, whereas, in North America, NetZero energy or NetZero GHG are more prevalent. Lack of agreement and notwithstanding definitions matter to generating useful design and construction solutions.

Jerry believes that the goal for achieving NetZero is a one way track. Either you have designed and operated a NetZero building, or you have not. Either way, agreement on definition and strategy has to be made during early design and confirmation of actual NetZero via measurement and verification is necessary for integrity and feedback into the design process.

Figure 58 Net Zero Carbon Buildings?

# Standardization versus customization

For Jerry, it is less of the perceived battle between customization and standardization and more of the cleverness of the engineers involved with the project. Great architecture and engineering combined with technology and software tools can solve just about any design issue.

The real construction issue going forwards, according to Jerry, is the lack of skilled labour, and this will lead to increased costs of labour which will, in turn, lead to more off-site construction and modular, factory-built homes.

The confluence of skills shortages, modular off-site construction methods and high technology should lead to a revolution in residential housing and, subsequently, building performance. This will be the future of green building!

Scan the QR code to visit
reinventingreenbuilding.com

# Reader Notes: The Godfather of Green

Readers are encouraged to note their favourite insights, triggered ideas, and next steps.

# Property Development 101

www.ingramcontent.com/pod-product-compliance
Lightning Source LLC
Chambersburg PA
CBHW042040200426
43209CB00060B/1708